Eagan Press Handbook Series

Dairy-Based Ingredients

Ramesh Chandan

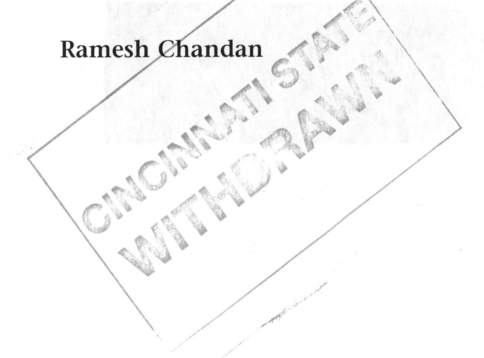

eagan® press
St. Paul, Minnesota, USA

Cover: Cow—Tri-Day Camero Fabulous, courtesy of Ann and
Curtis Day, Tri-Day Holsteins, Shippensburg, PA; plate heat
exchanger—courtesy of APV Heat Transfer; image of milk and
dairy products—©1997 Photodisc, Inc.

Library of Congress Catalog Card Number: 97-73710
International Standard Book Number: 0-913250-94-5

Printed in the United States of America on acid-free paper

American Association of Cereal Chemists
3340 Pilot Knob Road
St. Paul, Minnesota 55121-2097, USA

About the Eagan Press Handbook Series

The Eagan Press Handbook series was developed for food industry practitioners. It offers a practical approach to understanding the basics of food ingredients, applications, and processes—whether the reader is a research chemist wanting practical information compiled in a single source or a purchasing agent trying to understand product specifications. The handbook series is designed to reach a broad readership; the books are not limited to a single product category but rather serve professionals in all segments of the food processing industry and their allied suppliers.

In developing this series, Eagan Press recognized the need to fill the gap between the highly fragmented, theoretical, and often not readily available information in the scientific literature and the product-specific information available from suppliers. It enlisted experts in specific areas to contribute their expertise to the development and fruition of this series.

The content of the books has been prepared in a rigorous manner, including substantial peer review and editing, and is presented in a user friendly format with definitions of terms, examples, illustrations, and trouble-shooting tips. The result is a set of practical guides containing information useful to those involved in product development, production, testing, ingredient purchasing, engineering, and marketing aspects of the food industry.

Acknowledgment of Sponsor for *Dairy-Based Ingredients*

Eagan Press would like to thank the following company for its financial support of this handbook:

Davisco Foods International
Eden Prairie, Minnesota
800-757-7611

Eagan Press has designed this handbook series as practical guides serving the interests of the food industry as a whole rather than the individual interests of any single company. Nonetheless, corporate sponsorship has allowed these books to be more affordable for a wide audience.

Acknowledgments

The author thanks the following individuals for assisting, reviewing, and providing valuable counsel in the preparation of this book:

Gint Behrens, Land O'Lakes, Inc., Arden Hills, MN

Rulon Chappell, Chappell International, Inc., St. Paul, MN

Staff at Dairy Management Inc., Rosemont, IL

William Davidson, General Mills Inc., Minneapolis, MN

Todd Gusek, Cargill Inc., Minnetonka, MN

Loretta Kolberg, General Mills Inc., Minneapolis, MN

James Langler, General Mills Inc., Minneapolis, MN

Karen Schmidt, Kansas State University, Manhattan, KS

Glenn Van Hulle, General Mills Inc., Minneapolis, MN

Joseph Warthesen, University of Minnesota, St. Paul, MN

Contents

Properties of Milk and Its Components

Milk and dairy-based ingredients are used as components of many food products. Their contributions consist of unique flavor, desirable texture, excellent nutritive value, and a widely accepted "natural" image. In many instances, the success of the product in the marketplace is significantly enhanced by incorporation of traditional functional ingredients familiar to the consumer. Thus, dairy ingredients provide a consumer-friendly label on packaged foods.

Dairy ingredients contribute a number of characteristics critical to a food product. These include the emulsifying and stabilizing ability of *caseinates,* the gelling properties of *whey* protein concentrates and isolates, the water-absorption capacity of high-heat nonfat dry milk, and the browning of *lactose* during heat processing. Furthermore, the crystallization characteristics of lactose and the hydrolytic activity of the enzyme *lactase* are important in confectionery and frozen products. In addition, butter flavor carryover can be achieved with enzyme-modified butterfat and various cheese flavors imparted by enzyme-modified cheeses. Therefore, a food developer can select an appropriate dairy-based ingredient to create certain desirable attributes in foods. An understanding of the functional properties of dairy ingredients allows food technologists to utilize their potential contribution to product characteristics to meet consumer expectations.

Milk Composition

Milk may be defined various ways. Chemically speaking, milk is a complex fluid in which more than 100 separate chemical compounds have been found. Its major components are water, fat, lactose, *casein,* whey proteins, and minerals (or *ash*) in amounts varying with the milk of various species of animals. However, for any given species, the range of values for the constituents of milk is fairly constant.

From a physiological standpoint, milk is the secretion of the normally functioning mammary gland of the females of all mammals, which is produced for some time following *parturition* for the nourishment of the young of the species during the initial period of growth.

In terms of physical chemistry, milk is an opaque, whitish fluid of multidisperse phases. The true solution contains lactose, vitamins,

---

In This Chapter:

Milk Composition

Constituents of Milk
  Major Constituents
  Minor and Trace
    Constituents

Physical Properties
of Milk
  Color
  Flavor
  Density and Specific
    Gravity
  Surface Tension
  Foaming
  Viscosity
  Specific Heat
  Electrical
    Conductivity
  Freezing Point
  Boiling Point
  Refractivity

Caseinates—Compounds derived by the interaction of alkali with **casein**, the major milk protein.

Whey—The watery liquid remaining after the curd is formed in the manufacture of cheese and fermented or acidified dairy products.

Lactose—Milk sugar, composed of glucose and galactose.

Lactase—The enzyme that splits lactose (milk sugar) into glucose and galactose.

Ash—The residue left when a substance is incinerated at a very high temperature for analysis.

Parturition—The act or process of giving birth.

Colloidal phase—The portion of milk containing dispersed particles ranging in diameter from 10^{-5} to 10^{-7} cm.

Emulsion—A homogeneous dispersion of two dissimilar immiscible liquid phases. If oil is dispersed in water, it is an oil-in-water (O/W) emulsion. If water is dispersed in oil, it is a water-in-oil (W/O) emulsion.

Low-fat milk—Milk containing at least 8.25% solids-not-fat and with fat reduced to deliver not more than 3 g of milkfat per serving of 8 fl oz. Also termed *light milk*.

Nonfat milk— Milk containing at least 8.25% solids-not-fat and with fat reduced to deliver not more than 0.5 g of milkfat per serving of 8 fl oz. Also termed *fat-free* or *skim milk*.

Colostrum—The first milk secreted by an animal just before and after the birth of its young.

Oestrum/Estrus—Period of sexual receptivity (heat) in female mammals.

Casein micelles—Large colloidal particles that are complexes of protein and salt ions, principally calcium and phosphorus.

acids, enzymes, and some inorganic salts. The *colloidal phase* contains casein, calcium phosphate, and globular proteins. Fat exists in the form of an oil-in-water type of *emulsion,* with fat globules varying from 0.1 to 22 μm in diameter.

As a food ingredient or consumed by itself, milk provides an excellent nutritional profile in the human diet. Nutrition experts consider milk an exceptionally complete food because it contains significant levels of required nutrients such as protein, fat, carbohydrates, minerals, and several vitamins. *Low-fat* and *nonfat* milks are increasingly popular in fat-reduced and fat-free food formulations.

Worldwide, milk of the cow is by far of more commercial importance than milk of any other mammal. In the United States, the term "milk" legally refers to cow's milk. Milk from other species is labeled to indicate the type: sheep's milk, goat's milk, etc. Milk is the whole, clean lacteal secretion of one or more healthy cows, properly fed and kept, excluding that obtained within 15 days before calving and three to five days after. *Colostrum,* the milk secreted immediately after giving birth, is not considered milk from a legal standpoint. The U.S. Public Health Service's definition of Grade A milk is "the lacteal secretion practically free from colostrum, obtained by complete milking of one or more healthy cows, which contains not less than 8.25% milk solids-not-fat (MSNF) and not less than 3.25% milkfat."

Constituents of Milk

Milk is composed of water, milkfat, and MSNF. The MSNF consists of protein, lactose, and minerals. These solids are also referred to as *skim solids*, or *serum solids*. The term *total solids* refers to the serum solids plus the milkfat.

The major components of commercial raw milk (1–3) are illustrated in Figure 1-1. On a dry basis, the composition of milk solids is shown in Figure 1-2. The composition of the MSNF portion of milk is given in Figure 1-3.

MAJOR CONSTITUENTS

The major constituents of milk vary more widely in individual cow's milk than in pooled market milk. Factors affecting the milk such as breed of cow, intervals of milking, stages of milking, differ-

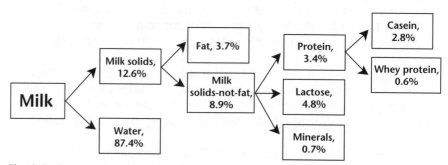

Fig. 1-1. Gross composition of milk, showing major constituents.

ent quarters of udder, lactation period, season, feed, nutritional level, environmental temperature, health status, age, weather, *oestrum,* gestation period, and exercise are known to cause variations in fat, protein, lactose, and mineral levels in milk derived from individual cows (4,5). In general, these variations tend to average out but display an interesting seasonal pattern in commercial milk used by food processors, which may have an important impact on properties of the finished products. An approximately 10% variation in fat and protein is observed in milk received in July and August (lowest level) compared to that received in October and November (highest level). Variations in protein and fat during the year affect yogurt and cheese manufacture and whey protein production. For example, appropriate adjustments, such as fortification with additional nonfat milk solids, are necessary in yogurt production to ensure uniform viscosity throughout the year.

Milk proteins. Caseins make up approximately 80% of milk proteins. The remaining proteins are classified as whey proteins. Milk proteins and their concentration in milk are shown in Table 1-1.

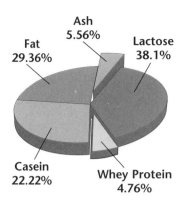

Fig. 1-2. Composition of milk solids of whole milk.

TABLE 1-1. Milk Proteins [a]

Type	Nomenclature	Concentration (g/L of milk)
Casein		24–28
	α_{s1}-Casein	12–15
	α_{s2}-Casein	3–4
	β-Casein	9–11
	κ-Casein	3–4
	γ-Casein	1–2
Whey protein		5–7
	β-Lactoglobulin	2–4
	α-Lactalbumin	1–1.5
	Bovine serum albumin	0.1–0.4
	Immunoglobulins	0.6–1.0
	Proteose peptones	0.6–1.8

[a] From: Functional Properties of Milk Proteins, by O. Robin, S. Turgeon, and P. Paquin, in: *Dairy Science and Technology Handbook,* Vol. 1, Y. H. Hui, Ed. ©1993 by VCH Publishers. Used by permission of John Wiley & Sons, Inc.

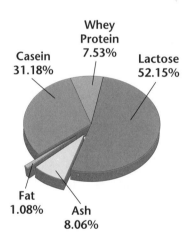

Fig. 1-3. Composition of nonfat solids of skim milk.

Besides having a biological and nutritional role, caseins and caseinates are important because of their structure, charge, and physical properties. Caseins become insoluble when the milk is acidified and the pH is reduced to 4.6, while the whey proteins remain in solution. Caseins exist in milk as particles called *micelles,* which are made up of calcium phosphate and casein complexes. The micelles are spherical particles varying in size from 50 to 200 nm and containing thousands of protein molecules. Caseins are further divided into α_{s1}, α_{s1}, β, and κ fractions. A γ-fraction is derived from the breakdown of β-casein by inherent proteolytic enzymes of milk. Each of

Rennin—A milk coagulatory enzyme found in the gastric juice of the fourth stomach of calves.

Serum phase—The watery portion of a fluid; the portion of milk without fat globules and casein micelles.

Lipids—A class of compounds found in nature that are soluble in organic solvents such as ether or hexane. Examples are triacylglycerols, cholesterol, and vitamin A.

the casein fractions exhibits distinct chemical properties. For example, α_{s1} and α_{s2} caseins have eight and 10–13 phosphoserine units, respectively. β-Casein has five phosphoserine units. All the casein fractions except κ-casein are precipitated by calcium, whereas κ-casein, which contains one phosphoserine unit, is not calcium-sensitive. Only κ-casein contains a carbohydrate moiety. Calcium (30 mM) and phosphate (20 mM) are complexed with α_{s1}, α_{s2}, and β-caseins, whereas κ-casein stabilizes the colloidal particles (micelles) by surface binding. During cheesemaking, the stabilizing κ-casein is cleaved by the enzyme *rennin,* resulting in destabilization of the micelle and subsequent curd formation.

Whey proteins are located in solution in the *serum phase* of milk along with minor proteins and enzymes, including lactoperoxidase, lactotransferrin, lysozyme, glycoprotein, serum transferrin, and degradation products derived from casein.

Caseins and whey proteins are distinguishable from each other by their physical and chemical characteristics (Table 1-2). Caseins have a distinct, disordered molecular structure that lacks stabilizing disulfide bridges. This characteristic structure makes the casein fraction precipitate in acidic conditions as well as in the presence of di- and polyvalent ions of various salts. Casein molecules exist naturally in an open and extended state. Therefore, heat has little or no effect on them. However, severe heat treatment results in detachment of phosphate groups as well as formation of brown pigments by interaction with milk sugar (the Maillard reaction).

Whey proteins have a relatively more ordered molecular structure, which contains disulfide linkages. Accordingly, they are not vulnerable to precipitation under acidic conditions or by polyvalent ions. Like other globular proteins, they can be heat-denatured, resulting in gel formation. β-Lactoglobulin complexes with κ-casein in milk subjected to rigorous heat treatment.

In food systems, milk proteins contribute to properties of the final product. Table 1-3 summarizes the major functional characteristics of milk proteins.

Milkfat. The fat in milk occurs in microscopic globules in an oil-in-water type of emulsion. The globules vary in size from 0.1 to 22 μm in diameter. The *lipid* content of milkfat is 97–98% triacylglycerols, 0.2–1% phospholipids, 0.2–0.4% sterols, and traces of fatty acids, as well as vitamins A, D, E, and K. Milkfat is made up of 65% saturated,

TABLE 1-2. Major Physical and Chemical Differences in Milk Proteins

Caseins	Whey Proteins
Strong hydrophobic regions	Both hydrophobic and hydrophilic regions
Little cysteine content	Both cysteine and cystine present
Random coil structure	Globular structure with helical contents
Heat stable	Easily heat denatured and insolubilized
Precipitate in acidic conditions and insoluble at pH 4.6	Stable in mild acidic environment
Precipitated by di- and polyvalent ions	

TABLE 1-3. Functional Characteristics of Milk Proteins

Functionality	Casein or Caseinates	Whey Proteins
Hydration, water binding	Very high, minimum at pH 4.6	Water-binding capacity increases with denaturation of the protein
Solubility	Insoluble at pH 4.6	Soluble at all pH levels. If denatured, insoluble at pH 5
Viscosity	High at or above pH 6	Low for native protein. Higher if denatured
Gelation	No thermal gelation except in the presence of Ca^{+2}. Micelles gel with rennin	Heat gelation at 70°C (158°F) or higher; influenced by pH and salts
Emulsifying ability	Excellent at neutral and basic pH	Good except at pH 4–5, if heat denatured
Foam formation	Good overrun. κ-Casein best, followed by β- and $α_{s1}$-caseins. Poor foam stability	Good overruns. β-Lactoglobulin better than α-lactalbumin
Flavor binding	Good	Retention varies with degree of denaturation

32% monounsaturated, and 3% polyunsaturated fatty acids. It contains 7% short-chain fatty acids (C4–C8), 15–20% medium-chain fatty acids (C10–C14), and 73–78% long-chain fatty acids (C16 or higher). The cholesterol content of whole milk (3.3% fat) and skim milk is 14 mg/100 ml and 2 mg/100 ml, respectively. The functional properties of milkfat are attributed to its fatty acid make-up.

Milk sugar. Lactose is the major carbohydrate of milk, occurring at a 4.5–4.9% level. It consists of two forms (α and β) that differ in their sweetness and solubility. The α-form is less soluble (70 g/L at 15°C) than the β-form. Crystallization is important in the manufacture and utilization of several dairy ingredients. An equilibrium mixture of α- and β-lactose, formed by mutarotation, exhibits a solubility of 170 g/L of water. β-Lactose is slightly sweeter than the α-form. Overall, in sweetening power, lactose is only 16–33% as sweet as sucrose. The sweetening power is concentration dependent.

MINOR AND TRACE CONSTITUENTS

The typical composition of major milk minerals is shown in Table 1-4.

Mineral concentration in milk is related to physical-chemical equilibria, which are important in processing, nutritive value, and shelf life of dairy products. Minerals include chloride, PO_4^{2+}, and citrates of K^+, Na^+, Ca^{2+}, and Mg^{2+}. Their concentration is <1% in milk, but they are involved in heat stability and alcohol coagulation of milk, age-thickening of sweetened condensed milk, *feathering* of coffee cream, rennin coagulation, and clumping of fat globules upon *homoge-*

Feathering—The limited coagulation of milk proteins when cream is added to hot liquid (e.g., coffee), characterized by the appearance of small particles on the surface.

Homogenization—A process for reducing the size of milkfat globules in milk. Upon undisturbed storage at 7°C, homogenized milk shows no visible cream separation.

TABLE 1-4. Major Mineral Composition of Milk

Milk Salt Constituents	Concentration (mg/100 ml whole milk)	Concentration Range (mg/100 ml)
Calcium	123	110–130
Magnesium	12	9–14
Phosphorus (total)	95	90–100
Phosphorus (inorganic)	75	70–80
Sodium	58	35–90
Potassium	141	110–170
Chloride	99	90–110
Sulfate	10	...
Carbonate (as CO_2)	20	...
Citric acid	160	...

Ultrafiltration—A process that uses a semipermeable membrane to separate fractions based on molecular size.

Electrodialysis—A process that uses electric charge to separate substances in solution (in this case, removing minerals from whey or milk fractions).

TABLE 1–5. Trace Minerals of Milk

Constituent	Concentration (μg/100 ml of milk)
Iron	30–60
Zinc	200–600
Copper	10–60
Manganese	2–5
Iodine	2–6
Fluoride	3–22
Selenium	0.5–6.7
Cobalt	0.05–0.13
Chromium	0.8–1.3
Molybdenum	1.8–12
Nickel	0–5
Silicon	75–700
Vanadium	0–31
Tin	4–50
Arsenic	2–6

nization. The calcium level of milk influences the firmness of curd during cheesemaking.

Ash, the white residue after incineration of a given weight of milk, is used as a measure of the mineral content of milk. It is not identical to milk mineral level because of decomposition and volatilization of certain minerals due to heat. Ash contains carbonates derived from organic constituents; sulfates from proteins; phosphate, partly from casein, which may contain approximately 1.62% phosphate; and chloride, which is partly lost (45–50%). Citric acid is completely lost. The average ash content is 0.70%, which is equivalent to 0.90% minerals.

All the minerals considered essential for human nutrition are found in milk. They exist in milk in different states. Sodium, potassium, and chloride are ionic forms and in true solution. They permeate freely across the membrane during *ultrafiltration* and *electrodialysis* of milk and whey. Calcium, magnesium, inorganic phosphorus, and citrate exist in both colloidal and diffusible forms, depending on the pH of milk. Approximately 20–30% of diffusible calcium and magnesium exists as free ions and the remainder as salts of citrate and phosphate. As the pH of milk drops, the colloidal form is converted progressively to the ionic form. At pH 4.4, most of the minerals are in diffusible form.

By lowering the pH of milk or whey, protein concentrates of low mineral content and different mineral ratios can be produced.

Trace elements are those constituents found in the parts per million level in milk. Table 1-5 shows their levels.

In addition, several nonprotein nitrogen compounds, vitamins, and some organic materials are present in milk (7, Table 1-6).

Physical Properties of Milk

COLOR

The color of milk is perceived by consumers to be indicative of purity and richness. Its white color is due to the scattering of reflected light by the inherent ultramicroscopic particles, fat globules, colloidal casein micelles, and calcium phosphate (5). The intensity of white color is directly proportional to the size and number of particles in suspension. Homogenization increases the surface area of fat globules significantly as a result of the breakup of larger globules. Accordingly, homogenized milk and cream are whiter than their unhomogenized counterparts. Lack of fat globules gives skim milk a blue tinge.

Cow's milk contains the pigments carotene and xanthophyll, which tend to give a golden yellow color to the milkfat.

FLAVOR

The flavor of milk is a property difficult to define, but there is no doubt that taste and aroma are critical to the assessment of milk. Flavor constitutes a critical criterion of quality for the consumer. It is a sensory property in which odor and taste interact. The sweet taste of lactose is balanced against the salty taste of chloride, and both are somewhat moderated by proteins. This balance is maintained over a fairly wide range of milk composition even when the chloride ion level varies from 0.06 to 0.12%. Saltiness can be detected by sensory tests in samples containing 0.12% or more of chloride ions and becomes marked in samples containing 0.15%. Some workers attribute the characteristic rich flavor of dairy products to the lactones, methylketones, certain aldehydes, dimethyl sulfide, and certain short-chain fatty acids.

Although milk has a clean, pleasantly sweet flavor, it is quite bland, and therefore any off-flavors are readily discernible. Off-flavors result when the balance of flavor compounds is altered by microbiological action, dairy farm or processing conditions, or chemical or biochemical reactions. Appendix A describes off-flavors and gives potential causes.

DENSITY AND SPECIFIC GRAVITY

The *density* of milk with 3–5% fat averages 1.032 g/cm³. Accordingly, the weight of 1 L of milk is 1.03 kg. To convert the weight of milk from kilograms per liter to pounds per gallon, the number is multiplied by 8.34.

The average *specific gravity* of milk at 15.5°C (60°F) is 1.032. It varies from 1.028 to 1.035 in commercial milk. The specific gravities of milkfat (0.93), MSNF (1.62), and water (1.0) determine the specific gravity of milk. Specific gravity is increased by the removal of fat and lowered by the addition of water.

Density—Mass per unit volume.

Specific gravity—Ratio of the density of a product and the density of water at the same temperature.

TABLE 1–6. Other Trace Components of Milk [a]

Constituent	Concentration per 100 ml of Milk
Vitamins	
A	40 µg retinol equivalent
B	
B₁	45 µg
B₂	175 µg
Niacin	90 µg
B₆	50 µg
Pantothenic acid	350 µg
Biotin	3.5 µg
Folic acid	5.5 µg
B₁₂	0.45 µg
C	2 mg
D	4 IU
E	100 µg
K	5 µg
Nonprotein nitrogen compounds (NPN)	
Total NPN	22.9–30.8 mg
Urea N	8.4–13.4 mg
Creatine N	0.6–2 mg
Uric acid N	0.5–0.8 mg
Orotic acid N	1.2–1.3 mg
Choline	4.3–28.5 mg
N-acetylneuraminic acid	12–27 mg
Miscellaneous	
Lactic acid	3.4–10.4 mg
Acetic acid	0.3–5 mg
Formic acid	1–8.5 mg

[a] From: Chemistry and Physics, by H. D. Goff and A. R. Hill, in: *Dairy Science and Technology Handbook,* Vol. 1, Y. H. Hui, Ed. ©1993 by VCH Publishers. Used by permission of John Wiley & Sons, Inc.

Surface tension—Forces causing a reduction in surface area, which is a characteristic property of a liquid.

Protease—An enzyme that attacks and hydrolyzes proteins.

Viscosity—Resistance to flow. A measure of the friction between molecules as they slide past one another.

SURFACE TENSION

Surface properties are involved in adsorption phenomena and the formation and stability of emulsions. They are relevant to creaming, fat globule membrane function, foaming, and emulsifier use in dairy products. Normal cow's milk has an inherent surface activity. Its surface tension approximates 70% of that of water. The *surface tension* of whole milk is usually about 50–52 mN/m (or dyn/cm) at 20°C. The surface tension of skim milk is 55–60 mN/m. For cream, it is approximately 46–47 mN/m (8,9). Casein, along with the proteolysis products *protease*-peptones, is largely responsible for the surface activity. Whey proteins make little contribution. Fat reduces surface tension by a physical effect. Lactose and most of the salts tend to raise it when they are present in true solution.

Surface tension decreases as milk temperature rises. Processing treatments such as heating, homogenization, and shear tend to increase surface tension.

FOAMING

The formation of stable foam depends upon two main factors. First, the lowering of the surface tension allows the gathering and spreading of the surface-active components into thin films. Second, the films must be sufficiently elastic and stable to prevent the coalescence of the gas cells. A stable foam is thus formed when the surface tension of the liquid is not great enough to withdraw the film from between the gas cells and when the stabilizing agent has great internal viscosity.

Foaming of milk is at a minimum at 30–35°C. At 60°C, the foam volume is independent of the fat content. Below 20°C and above 30°C, the foaming tendency appears to increase. Fat tends to stabilize the foam formed below 20°C, for instance, during churning. Skim milk produces slightly more stable foam above 30°C than whole milk or light cream.

Foaming properties affect handling of milk products and how dairy-based ingredients are incorporated into other products.

VISCOSITY

Whole milk and skim milk display *viscosities* of 2.0–2.1 and 1.5–1.8 cP (or mPa/sec) at 20°C, respectively (4,8). Whey has a viscosity of 1.2 cP. The viscosity of milk and cream creates the impression of "richness" to the consumer. From an organoleptic standpoint, viscosity contributes to mouthfeel and flavor release.

The casein micelles of milk contribute more to the viscosity of milk than any other constituent. Viscosity varies not only with changes in the physical nature of fat but also with the hydration of proteins. Alterations in the size of any dispersed constituents result in viscosity changes. The fat contributes less than casein but more than whey proteins. When fat globules are greatly subdivided by homogenization, an increase in viscosity is observed. The viscosity of skim milk decreases on heating to 62°C, after which it increases, apparently due to changes in protein hydration. An increase of temperature causes a marked reduction of viscosity. For example, at 20°C, milk is about half as viscous as at 0°C and at 40°C is approximately one-third of the value at 0°C.

SPECIFIC HEAT

The *specific heat* of milk products is a function of their composition. The values for whole milk, skim milk, 40% cream, butter, and whey at 15°C are 0.93, 0.95, 0.68, 0.53, and 0.97 BTU/lb·°F, respectively (3.89, 3.97, 3.35, 2.21, and 4.06 kJ/kg·K, respectively).

ELECTRICAL CONDUCTIVITY

Current passes through the milk by virtue of the activity of its ionic mineral constituents, of which the chloride ions carry 60–68% of the current. There is therefore a close correlation between the *electrical conductivity* of milk and its chloride content. The electrical conductivity of normal milk corresponds to that of approximately 0.25% sodium chloride solution (w/w) and ranges from 45 to 55 A·v^{-1}m^{-1} (8) or 45–55 x 10^{-4} mho (1). Conductivity of milk is the basis of the new ohmic process for sterilizing milk.

Lactic acid accumulates as a result of fermentation during the manufacture of yogurt and fermented dairy products, converting calcium and magnesium to ionic form and thereby increasing the conductivity reading. Thus, the progress of fermentation can be followed by increases in the conductivity of the yogurt base. Also, demineralization of whey and its fractions, leading to loss of ionic minerals, is monitored using a conductivity meter.

FREEZING POINT

The freezing point of milk is generally expressed as degrees Hortvet (H), reflecting the commonly used Hortvet crysoscope procedure. Determination of freezing point is a parameter widely used in the industry for detection of adulteration of milk with water. The freezing point of milk has a relatively narrow range (from –0.520 to –0.560°C, the average being –0.540°C). It corresponds to the freezing point of an 0.85% sodium chloride solution (w/w). Addition of water raises the freezing point, and readings above –0.520 support strong suspicion of watering the milk. As little as 3% water added to milk can be detected by this method.

Lactose and chloride are the major milk constituents responsible for 70–80% of the overall depression in the freezing point of milk.

BOILING POINT

A solution boils at a higher temperature than does the pure solvent, according to the concentration of the dissolved substance. The boiling point of milk is 100.17°C. The milk constituents in true solution are mainly responsible for the elevation of the boiling point above 100°C. Elevation of the boiling point is based on the same

Specific heat—Number of calories required to raise the temperature of 1 g of a substance by 1 degree C.

Electrical conductivity—The reciprocal of electrical resistance exhibited by a 1-cm cube of conductor (solution containing electrolytes).

Refractive index—A physical property of a substance that relates to how light is refracted from the material. Usually used to indirectly measure some other property such as concentration.

principles as depression of freezing point. However, for detecting added water, the freezing point method is far superior on the grounds of accuracy and convenience.

REFRACTIVITY

The refraction of light by a solution is a function of the molecular concentration of the solute in solution. Each solute maintains its own refractivity, and the *refractive index* of a mixture is that of the total of the refractive indices of the substances plus that of the solvent. The components of milk contributing to its refractive index in descending order of importance are water, proteins, lactose, and minor constituents. Whey proteins are more important than casein. The refractive index of milk at 20°C is 1.3440–1.3485.

The refractive index of butterfat is distinct enough to indicate its authenticity.

References

1. National Dairy Council. 1993. *Newer Knowledge of Milk and Other Fluid Dairy Products.* The Council, Rosemont, IL.
2. Varnum, A. H., and Sutherland, J. P. 1994. *Milk and Milk Products.* Chapman & Hall, New York.
3. Harper, W. J., and Hall, C. W. 1976. *Dairy Technology and Engineering.* Avi Publishing Co., Westport, CT.
4. Wong, N. P., Jenness, R., Keeney, M., and Marth, E. H., Eds. 1988. *Fundamentals of Dairy Chemistry,* 3rd ed. Van Nostrand Reinhold, New York.
5. Goff, H. D., and Hill, A. R. 1993. Chemistry and physics. In: *Dairy Science and Technology Handbook,* Vol. 1. Y. H. Hui, Ed. VCH Publishers, New York.
6. Robin, O., Turgeon, S., and Paquin, P. 1993. Functional properties of milk proteins. In: *Dairy Science and Technology Handbook,* Vol. 1. Y. H. Hui, Ed. VCH Publishers, New York.
7. Riel, R. 1985. Composition and physicochemical structure of milk. In: *Dairy Science and Technology, Principles and Application.* Les Presses de L' Universite Laval, Quebec, Canada.
8. Walstra, P., and Jenness, R. 1984. *Dairy Chemistry and Physics.* John Wiley, New York.
9. Singh, H., McCarthy, O. J., and Lucey, J. A. 1997. Physicochemical properties of milk. In: *Advanced Dairy Chemistry,* Vol. 3. P. F. Fox, Ed. Chapman & Hall, New York.

Basic Milk Processing

Whole milk, low-fat milk, and skim milk are rarely used as ingredients in food products other than dairy products because of their excessive moisture content (88%) and the possible undesirable interactions among food constituents brought about by the thermal treatments often necessary during processing. For example, lactose may become insoluble, causing grittiness, or *reducing sugars* in a food system may react with milk proteins to cause browning discoloration. Nevertheless, in certain instances, fluid milk may be the ingredient of choice for economic reasons. The food formulator may be able to adjust the quantity of water in a food by compensating for water contained in fluid milk.

Fluid milk, in forms ranging from whole milk to skim milk, is the main ingredient purchased by dairy processors and manufacturers of yogurt and other grade A fermented milk products. In addition, cheese plants and frozen dessert manufacturers purchase milk (manufacturing grade and Grade A) in bulk quantities.

Specifications for fluid whole milk and skim milk are shown in Table 2-1. In most states, milk is designated as Grade A, manufacturing grade, or reject. Grades A, B, or C are also used by some agencies. More than 95% of all the milk produced now conforms to Grade A requirements as defined in the Pasteurized Milk Ordinance (PMO) (1, Table 2-2).

A typical fluid milk and dairy product operation is shown in Figure 2-1 (2–4).

Raw Milk Handling and Storage

Bulk milk handling is a key step in the handling of good-quality milk. Dairy farms produce milk under supervision by the U.S. Public Health Service. Several quality milk programs exist in the industry to encourage sanitary raw milk production. They entail control of temperature, sediment, microbial load, *somatic cell count*, off-flavors, and antibiotics in the milk supply.

A description of milk specifications is given in Box 2.1.

Virtually all the raw milk at the plant is delivered in tank trucks of 1,500- to 5,000-gal capacity. Unloading of milk involves agitation of the milk in the tanker truck, inspection for off-flavors, taking a representative sample, and pumping milk from the truck to a storage tank or silo.

Storage tanks may be refrigerated or insulated. They hold milk up

In This Chapter:

Raw Milk Handling and Storage

Separation

Standardization
  Fat Standardization
  Standardization of Fat and Solids-Not-Fat

Pasteurization

Homogenization

Packaging and Storage

Reducing sugar—A sugar molecule containing free aldehyde groups that are available to react with a free amino group of protein, peptide, or amino acid.

Somatic cell count—Count of the mixture of dead epithelial cells and leukocytes that migrate into milk from the udder.

TABLE 2-1. Specifications and Standards for Purchase of Raw Fluid Whole and Skim Milk

Standards (Chemical and Physical)	Milk	Skim Milk	Analytical Method[a]
Titratable acidity, % max.	0.16	0.16	AOAC/APHA
Freezing point, °C max.	−0.530	−0.530	Cryoscope, APHA
Fat range, %	3.25–3.70	0.04–0.10	Babcock/Gerber, AOAC
Total solids, %	12.0–13.0	8.9–9.3	AOAC
Antibiotics	Negative	Negative	Disk assay/Delvo Test, APHA
Temperature, °C (°F)	4.4–7.2 (40–45)	4.4–7.2 (40–45)	...
Flavor and odor objectionable taints	Clean, fresh, no objectionable taints	Clean, fresh, no objectionable taints	...
Microbiological standard			
Standard plate count, CFU[b]/ml max.	300,000[c] 1,000,000[d]	300,000[c] 1,000,000[d]	APHA
Direct microscopic clump count, CFU/ml max.	200,000	200,000	
Somatic cells count, per ml, max.	750,000	750,000	Direct microscopic somatic cell count, electronic somatic cell count, flow cytometry/opto-electronic somatic cell count, and membrane filter DNA/somatic cell count methods.
Listeria	Negative	Negative	APHA
Salmonella	Negative	Negative	APHA
Coliform, CFU/ml, max.	10	10	APHA
Psychrotrophic count, CFU/ml, max.	10	10	APHA

[a] AOAC = AOAC International, APHA = American Public Health Association.
[b] Colony-forming units.
[c] Grade A commingled milk; 100,000 for producer herd milk.
[d] Manufacturing Grade, commingled milk; 500,000 for producer herd milk.

Psychrotrophic—Describing cold-tolerant microorganisms capable of growing at 4–15°C (39–59°F).

to 72 hr (usually 24 hr) before processing. Silo tanks vary in capacity up to 60,000 gal. When emptied, they are cleaned within a maximum of 72 hr. Grade A milk for pasteurization must be stored at 1.6–4.4°C (35–40°F). The maximum bacterial count permitted is 300,000 colony-forming units (CFU) per milliliter as opposed to 100,000 CFU/ml at the farm. The higher count is justified because 1) pumping breaks the clumps of bacteria, giving higher counts, 2) there is more opportunity for contamination of milk as it comes in contact with equipment during handling and transfer, and 3) the longer time of storage allows more bacteria to grow. The issue at this point is the growth of bacteria (especially growth of *psychrotrophic* organisms) and the accompanying degradation of milk constituents, which gives rise to off-flavors and processing problems. In practice, most raw milk displays lower bacterial content, typically round 25,000 CFU/ml.

Separation

Use of a separator permits fractionation of whole milk into milk (or skim or low-fat milk) and cream. Fat globules are lighter (less dense) than the surrounding water phase and rise to the surface

TABLE 2-2. Various Pasteurized Milk Ordinance (PMO) Requirements for Grade A Compliance[a]

Standard	Raw Milk	Pasteurized Milk and Bulk-Shipped Heat-Treated Milk	Aseptically Processed Milk
Temperature	Cooled to 7°C (45°F) or less within 2 hr after milking, provided that the blend temperature after the first and subsequent milkings does not exceed 10°C (50°F)	Cooled to 7°C (45°F) or less and maintained thereat	...
Bacterial limits			
Standard plate count	Individual producer milk not to exceed 100,000 CFU[b]/ml prior to commingling with other producer milk. Not to exceed 300,000 CFU/ml as commingled milk prior to pasteurization	20,000 CFU/ml limit	No growth by tests specified in the PMO
Coliforms	...	Not to exceed 10 CFU/ml provided that, in case of bulk milk transport, tank shipments shall not exceed 100 CFU/ml	...
Somatic cell counts	Individual producer milk not to exceed 750,000 CFU/ml.	...	...
Drugs	No positive results on drug residue detection methods prescribed in the PMO	No positive results on drug residue detection methods for pasteurized milk as prescribed. Not applicable to cultured products	No positive results on drug residue detection methods for aseptically processed milk as prescribed. Not applicable to cultured products
Phosphatase	...	Less than 1 µg/ml by the Scharer Rapid Method. Less than 350 milliunits per liter for fluid products and less than 500 for other milk products by the Fluorometer or Charm ALP or equivalent	...

[a] Adapted from (1).
[b] Colony-forming units.

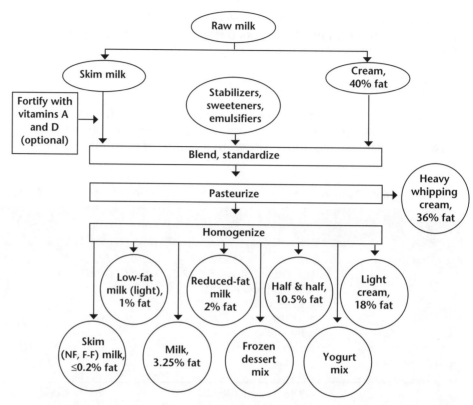

Fig. 2-1. Flow diagram for manufacture of fluid milk and other dairy products. Fortification of fluid milk products is optional for whole milk only. Stabilizers, sweeteners, and emulsifiers are blended only into mixes for yogurt and frozen dessert. NF = nonfat, F-F = fat-free.

when milk is left undisturbed. The separation of fat and water phases can be accelerated by the use of centrifugal forces. The operation of a separator is easily understood by Stoke's Law:

$$V = \frac{2r^2 (d_{\text{serum}} - d_{\text{fat}}) \bullet g}{9n}$$

where V = velocity of the fat globule, r = radius of the fat globule, d_{serum} = density of milk serum, d_{fat} = density of milk fat, g = the gravitational force, and n = viscosity of milk.

From the equation, it follows that V is directly proportional to g. If g is increased by centrifugation, fat globules can be separated in a relatively short time. Also, g is inversely proportional to the viscosity of the milk (n). As n decreases with increase in temperature, V increases. Therefore separation is more efficient at a warmer temperature, which keeps the fat in a liquid state. Skim milk should contain 0.05% fat or less if the separator is functioning properly.

Self-cleaning separators and cold milk separators are commonly used. The separator also permits the operator to predetermine the fat content. Increased back pressure on the cream discharge port increases the fat content in the milk being standardized.

If vitamins and minerals are to be added to fluid milk products, they are added at this point. Vitamins A and D are required for skim, low-fat, and reduced-fat milks and are optional for whole milk.

Standardization

Dairy products of varying compositions often must be mixed to ensure that a particular chemical composition is obtained or that particular standards are met. Various ingredients chosen by the manufacturer are blended in calculated quantities to yield a mix, which is used in another food product. This ensures that the composition of the product stays the same from batch to batch. *Standardization* of the mix may involve one or more variables of several ingredients of dairy or food origin.

Box 2-1. Specifications

A user plant purchases ingredients conforming to certain standards and specifications agreed upon with a supplier dairy plant. The specification sheet is a contract between the user and supplier. Any departure from the standard specification must be approved by the user. A specification sheet includes information related to regulatory issues, the legal definition, grade and kosher certification requirements, key processing steps related to ingredient functionality, and analytical procedures, as well as chemical, physical, and microbiological standards, packaging and shipping requirements, and storage conditions.

The following example illustrates these categories.

Ingredient name: Grade A Raw milk

Definition: Grade A milk conforming to Pasteurized Milk Ordinance of U.S. Public Health Service and Food and Drug Administration regulations. The product will be handled and processed in accordance with the standards recognized by Good Manufacturing Practices, Hazard Analysis Critical Control Points, and 3A Sanitary Standards of Interstate Milk Shippers.

Product requirement: The specifications for the product generally include chemical composition, titratable acidity, freezing point, fat, total solids, sensory, and microbiological standards.

FAT STANDARDIZATION

In the simplest case, a mixture with a specified fat content is produced from two ingredients containing different levels of fat. Use of the Pearson Square technique makes calculation of the needed amount of each ingredient fairly simple. Here is an example. A formulation calls for 1,000 lb of 1% low-fat milk; the processor has available skim milk containing 0.05% fat and cream containing 39.6% fat. To calculate the necessary amount of each, one draws a square or rectangle and enters at the left corners the fat percentages of skim milk and cream. The fat percentage of the finished product is written in the center.

Standardization—A step in dairy processing in which milkfat and/or solids-not-fat are made to conform to certain specifications by removal, addition, or concentration of milkfat.

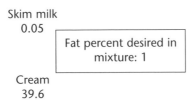

Skim milk
0.05

Fat percent desired in mixture: 1

Cream
39.6

Next, one subtracts diagonally across the rectangle, taking the smaller number from the larger one, and totals the two values.

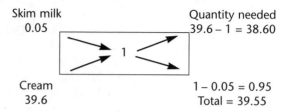

Skim milk
0.05

Quantity needed
39.6 – 1 = 38.60

1

Cream
39.6

1 – 0.05 = 0.95
Total = 39.55

The figures on the right side of the rectangle denote the ratios of the ingredients to blend in order to achieve the desired fat level in the finished product. Thus, when 38.6 lb of skim milk is blended with 0.95 lb of cream, one gets 39.55 lb of low-fat milk. To process 1,000 lb of low-fat milk, 976 lb of skim milk (38.6/39.55 x 1,000) and 24 lb of cream (0.95/39.55 x 1,000) are needed.

The calculation can be verified by calculating the fat contribution of each ingredient to the finished product.

Ingredient	Calculated quantity (lb)	Fat contribution (lb)
0.05% skim milk	976	0.5
39.6% cream	24	9.5
Total	1,000	10.0

Ten pounds of fat is therefore distributed in 1,000 lb of low-fat milk, which corresponds to 1% ([10/1,000] x 100) in the finished product.

STANDARDIZATION OF FAT AND SOLIDS-NOT-FAT

The algebraic method is also applicable to more complex standardization situations. In addition, the serum point method has been useful (5). Now, computers and appropriate software programs are common in dairy processing plants to determine formulations for products such as yogurt and frozen dessert, using multiple dairy and other ingredients, to make them conform to legal and quality standards.

Nonfat dry milk (NFDM) is an economical source of milk solids-not-fat in various dairy mixes. Unless it is dispersed correctly in the mix, it can cause lumps. To avoid processing problems, dry milk is reconstituted using specially designed recirculation equipment consisting of a vat (containing water or liquid milk) connected to a large, stainless steel hopper, funnel, or horn (containing dry ingredients). The flow of liquid from the vat is regulated by a two-way valve. A centrifugal pump is also connected to the hopper on the suction side. The pumped mixture is transported to the bottom wall of the vat. The blend is circulated until all the dry milk or other dry ingredients contained in the hopper are thoroughly dispersed. Figure 2-2 illustrates the principle.

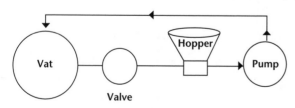

Fig. 2-2. Scheme for dispersion of dry milk in a mix.

In some plants, especially in frozen dessert plants, use of high-shear blenders for incorporation of dry ingredients is quite common. These blenders have agitators running at high speed, creating high-shear mixing.

The amount of NFDM needed to obtain a given solids level is dependent on its moisture content. For example, the weight of NFDM (4% moisture) needed to obtain 12% solids is 12/0.96 = 12.5%. To simulate skim milk, NFDM should be dispersed at the 9.2% level.

Pasteurization

The main purposes of heat treatment of milk are to render it safe for consumption and enhance its shelf life by inactivating most of the contaminating bacteria that may have gained entry during its production and handling. Heat treatment is an integral part of all processes used in dairy manufacturing plants.

In dairy processing, the term *pasteurization*, as defined by the PMO, is the process of heating every particle of milk or milk product, in properly designed and operated equipment, to one of the temperatures given in Table 2-3 and holding it continuously at or above that temperature for at least the corresponding specified time (1). *Ultrapasteurized* products are packaged in an aseptic atmosphere in presterilized containers and held refrigerated to achieve *extended shelf life*. Aseptically processed dairy products are those that have been subjected to sufficient heat processing and packaging in hermetically sealed containers to maintain commercial sterility of the product under normal nonrefrigerated storage conditions. *Ultrahigh temperature* (UHT) products are packaged aseptically in specially designed multilayer containers. They may be stored at room temperature for extended periods of time without the growth of bacteria.

Pasteurization—The process of heating milk and fluid dairy products to render them safe for human consumption by destroying disease-producing microorganisms. The process inactivates about 95% of the contaminating microorganisms.

Ultrapasteurization—The process of pasteurizing fluid dairy products by heating to 125–137.8°C (257–280°F) with a holding time of 2–5 sec to kill all pathogenic organisms, permitting storage at refrigerated temperature for an extended period.

Extended shelf life—Shelf life of 14–28 days at refrigerated temperature.

Ultrahigh temperature—A temperature of 135–150°C (275–302°F), used with a holding period of 4–15 sec. This sterilizes milk to permit storage at ambient temperature.

TABLE 2-3. Minimum Time-Temperature Requirements for Legal Pasteurization of Dairy Products

Process	Whole, Lowfat, or Skim Milk	Milk Products with Increased Viscosity, Added Sweetener, or Fat Content of 10% or More	Egg Nog, Frozen Dessert Mixes
Vat (batch)	30 min @ 62.7°C (145°F)	30 min @ 65.6°C (150°F)	30 min @ 68.3°C (155°F)
High-temperature short-time	15 sec @ 71.6°C (161°F)	15 sec @ 74.4°C (166°F)	25 sec @ 79.4°C (175°F) 15 sec @ 82.2°C (180°F)
Higher-heat shorter-time	1.0 sec @ 88.3°C (191°F) 0.5 sec @ 90°C (194°F) 0.1 sec @ 93.8°C (201°F) 0.05 sec @ 96.2°C (205°F) 0.01 sec @ 100°C (212°F)	Same	Same
Ultrapasteurized	2.0 sec @ 137.8°C (280°F)	Same	Same

TABLE 2-4. Heat-Treatment Processes for Milk Products

Process	Temperature	Holding Time	Purpose
Thermalization	145–149°F	15 sec	Preliminary heating
Vat pasteurization	145°F	30 min	Batch process
HTST[a] pasteurization	162–167°F	15 sec	Milk
HTST pasteurization	185–194°F	2–5 sec	Cream
Ultrapasteurization	257–280°F	2–5 sec	Cream products Refrigerated storage
UHT[b]	275–302°F	4–15 sec	Milk for ambient storage
Sterilization	240°F	20 min	Fluid canned products for ambient storage

[a] High-temperature short-time.
[b] Ultrahigh temperature.

The minimum temperature-time requirements for pasteurization are based on thermal death time studies on *Coxiella burnettii* (the Q fever organism), which is the most resistant pathogen that can be transmitted through milk.

Among milk proteins, caseins are relatively stable to heat effects. Whey proteins tend to *denature* progressively with the severity of heat treatment. They precipitate at 75–80°C, and 100% of the heat-denaturable proteins are denatured at 100°C. However, in the presence of casein in milk, denatured whey proteins complex with casein and no precipitation is observed.

Heat treatment may bring about physical and chemical changes in milk, depending on the severity of the temperature and duration of the treatment. Pasteurization results in minimal changes, but prolonged, intensive heat treatment can lead to changes in pH, whey protein denaturation, protein breakdown, and eventual coagulation of the milk. Heat also brings about interactions of certain amino acids with lactose, resulting in color changes in milk (*Maillard browning*), as observed in sterilized milk and evaporated milk products. These possess exceedingly cooked flavor and off-color. In general, pasteurized milk possesses the most acceptable flavor. Ultrapasteurized milk and UHT milk exhibit slightly cooked flavor. Viscosity and mouthfeel are also affected by heat treatment. Various heat treatment processes used in the food industry are shown in Table 2-4.

Shelf life of milk is a function of the microbial quality of raw milk, the temperature and time exposure during storage and handling, pasteurization conditions, equipment sanitation, packaging conditions, and subsequent distribution practices. UHT products have longer shelf life (under ambient storage conditions) than any other packaged fluid milk and cream products.

Types of pasteurizers are explained in Box 2-2.

Homogenization

Denaturation—The process that proteins undergo when subjected to certain physical or chemical treatments (e.g., heating) that cause disruption of the noncovalent bonds that maintain their secondary and tertiary structure. Denaturation causes profound changes in functional properties.

Maillard browning— Nonenzymatic browning caused by the heat-induced reaction of the ε-group of lysine with a reducing sugar. The bioavailability of lysine is lost, and characteristic flavor changes occur.

Creaming—The rising of milk-fat globules to the surface of milk left undisturbed, especially at cold temperatures, leading to a cream layer at the top of the milk.

Milkfat occurs in milk as fat globules varying in size from 0.1 to 22 μm. Their mean diameter is around 3.5 μm. Approximately 80–90% of globules are within the range of 2–6 μm. The number of globules ranges from 1.5 to 5 billion globules per milliliter of milk.

The objective of homogenization is to reduce the size of globules to 1 μm or less so that *creaming* is prevented in stored milk. Another effect of homogenization is relatively richer flavor due to an increase in surface area. Surface area is increased five times since one globule of 5 μm diameter gives rise to 125 globules of 1 μm diameter. Homogenization of milk also involves physical changes in milk proteins, resulting in lower curd tension and possibly increased digestibility due to faster coagulation in the stomach. Increased surface area makes homogenized milk more susceptible to *rancidity* and sunlight or UV-light-induced off-flavor. Homogenization creates an increase in viscosity in cream and ice cream, especially using high-pressure, low-temperature parameters.

According to the PMO (1), *homogenized* means that the milk or milk product has been treated to ensure breakup of the fat globules to such an extent that, after 48 hr of quiescent storage at 4.4°C (40°F), no visible cream separation occurs on the milk, and the fat percentage of the top 100 ml of milk in a quart, or of proportionate volumes in containers of other sizes, does not differ by more than 10% from the fat percentage of the remaining milk, as determined after thorough mixing.

Homogenization involves pumping milk at temperatures above 37.8°C (100°F) through a constriction or small orifice. Milk processing plants use two-stage homogenizers. The first stage subjects milk to 1,200–2,000 psi pressure, breaking the globules into much smaller globules as a result of shearing, shattering, and cavitation forces. The small globules tend to cluster, and the second stage (run at 500

Rancidity—An off-flavor caused by oxidation of fat or by the release of flavorful fatty acids from a triacylglycerol/triglyceride.

Box 2-2. Types of Pasteurizers

Indirect heat types. Vat pasteurization is now used only for specialty or by-products. It is not used in milk processing plants unless the plant output is rather small. The vat is a jacketed vessel for circulation of hot water, steam, cold water, or sweet water (iced water). It is equipped with an agitation device, a thermometer, a temperature-recording device, and a space heater to heat the area above the milk surface line 2.8 degrees C (5 degrees F) higher than the pasteurization temperature. Following a holding period of 30 min, the pasteurized product is cooled to 4.4°C (40°F) or less and subsequently packaged.

Most dairy plants now use a high-temperature short-time process. Continuous-type heat exchangers are of the plate type or tube type.

Plate heat exchangers (Fig. 2-3) consist of stainless plates held together in a frame. Mixing of thin channels of product and heating/cooling medium
continued on page 20

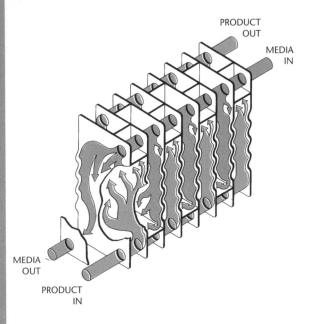

Fig. 2-3. A plate heat exchanger. (Courtesy APV Heat Transfer)

Box 2-2 continued from page 19

is prevented by separating plates with rubber gaskets. There are three sections of a plate heat exchanger. The first is the product-to-product regeneration section, involving flow of hot pasteurized milk (73.8°C, 165°F) in a direction opposite to that of incoming raw cold milk (4.4°C, 40°F). The hot milk transfers its heat to the cold milk, saving energy for both heating and cooling. About 80–90% efficiency of regenerators is achieved. Pasteurized milk exits this section at 20°C (68°F), while raw milk exits at 58.9°C (138°F). Second, the heating section elevates the temperature of warm raw milk from 58.9 to 73.8°C (138 to 165°F), using hot water and steam. The pasteurized milk at 73.8°C (165°F) enters a holding tube and is held for 16 sec. Pasteurized milk then enters a regeneration section where it is cooled to 20°C (68°F). Third, the cooling section chills the pasteurized milk from 20°C (68°F) to 1.6–4.4°C (35–40°F). A sweet water or cold glycol solution is circulated through the plates as a coolant.

Tube type heat exchangers are used for viscous products. The exchanger consists of two or three concentric tubes. In a triple tube system, the milk channel is sandwiched between two opposite-flowing heating- or cooling-medium channels. A double tube is useful for highly viscous products containing particulate matter. Product-to-product regeneration is also possible with double tubes.

Scraped surface or swept surface heat exchangers work better in the processing of viscous liquids. Scrapers constantly remove the product during heating or cooling to avoid stick-on or burn-on on the surfaces. Ice cream freezers, votators, cone vats, and process cheese kettles are examples of such exchanger types.

Direct steam heaters. These are used mostly in ultrapasteurized/ultrahigh temperature products.

The culinary (food grade) steam may be either directly injected into the milk stream, or, in a steam infusion system, steam heats a chamber of a vessel to 132.2–148.9°C (270–300°F) under pressure. Milk, as a thin layer or film, is introduced into the chamber, where it is heated almost instantaneously. The holding period for milk is usually 2–5 sec, after which the product is cooled and stored in aseptic tanks before packaging.

Since steam condensate increases the volume of fluid milk or cream, direct steam heaters are equipped with vaporizing units so that there is no dilution or concentration of the product. Milk at 71.6–73.8°C (161–165°F) (after the flow diversion valve) is treated with live culinary steam to raise its temperature to 76.7–115.6°C (170–240°F). The superheated milk then enters a vacuum chamber (20-in. vacuum) as a thin layer. Water vapor is continuously removed by condensation to maintain the original milk composition.

psi) breaks the clusters into individual globules. Homogenization and pasteurization are conducted in tandem, and either may be done first. Raw homogenized milk develops hydrolytic rancidity very rapidly because of activation of the inherent milk *lipase* system. Therefore, immediate pasteurization is necessary to inactivate the enzymes.

Lipase—An enzyme that hydrolyzes acyglycerols/ glycerides.

Packaging and Storage

The fluid product is held at 3.3–4.4°C (38–40°F) and subsequently packaged. Pasteurized dairy products are packaged in plastic blow-molded containers, gable-top paperboard containers, and plastic bag-in-box containers.

Aseptic packaging, using hydrogen peroxide as a sterilant, and sterile atmosphere (closed to outside air) is used for products with extended shelf life. The packaging materials include a combination of aluminum foil, saran, paper, polyvinylchloride, polypropylene, and polystyrene. Barriers to oxygen and moisture transmission are formed by coextrusion and lamination of these materials. The cost of the packaging depends on the shelf life desired. A high-cost oxygen barrier can maintain the quality of milk for up to four to six months.

A fluid dairy ingredient for use in a food plant may be custom-processed and packaged by a dairy. Butterfat content, milk solids-not-fat level, and certain processing conditions may be specified. It is transported at 3.3–4.4°C (38–40°F) in appropriate package size or in bulk to the user plant.

References

1. Department of Health and Human Services, Public Health Service, 1995. Grade "A" Pasteurized Milk Ordinance, revised. U.S. Government Printing Office, Washington, DC.
2. Robinson, R. K. 1994. *Modern Dairy Technology,* Vol. 1. *Advances in Milk Processing,* 2nd ed. Chapman & Hall, New York.
3. Harper, W. J., and Hall, C. W. 1976. *Dairy Technology and Engineering.* Avi Publishing Co., Westport, CT.
4. Rosenthal, I. 1991. *Milk and Dairy Products.* VCH Publishers, Inc., New York.
5. Marshall, R. T., and Arbuckle, W. S. 1996. *Ice Cream,* 5th ed. Chapman & Hall, New York.

Production and Specifications of Milk Concentrates

Dairy ingredients used in the formulation of various food products constitute milk in fluid, condensed, or dry form, which provide the desirable attributes of nutrition, water binding, fat holding, emulsification, viscosity, *gelation*, and foaming, as well as textural and flavor attributes. In addition, custom-made mixes may be fabricated by dairy plants for food plants producing yogurt, ice cream, and confectionery products. Also, milk and whey are fractionated to concentrate protein, fat, or mineral constituents to enhance their utility in food product performance. The typical composition of dairy ingredients is shown in Appendix B.

In general, the functional properties of a dairy ingredient are related to its chemical composition and the specific processing conditions to which it is subjected to modify its performance in a given food system. Selection of a dairy ingredient is largely based on the desired contribution of functional proteins, fats, lactose, and minerals in a given food. Cost and availability also contribute to the use of a particular ingredient. Recent trends in production of major dairy products (1) have had an impact on their cost and availability. In 1995, 36% of the milkfat in the United States was utilized in fluid milk and cream, and 34% was used in the production of cheese (Fig. 3-1A). Sales of nonfat dry milk (NFDM) are affected by supply and demand factors, which are influenced by government programs. NFDM uses are illustrated in Figure 3-1B. The end-uses of dry whole milk and dry buttermilk are illustrated in Figure 3-1C and D, respectively.

Effective utilization of milkfat has been a challenge for many years in view of its saturated *fatty acid* makeup and the subsequent controversial link to human cardiovascular disease. More recently, the butterfat surplus has disappeared, mostly because of reduced pricing of the fat.

Concentrated Milk Products

A series of dairy ingredients is obtained by removal of water from milk, low-fat milk, and skim milk (Fig. 3-2). Several procedures are

In This Chapter:

Concentrated Milk Products
  Condensed Products
  Dry Milk Products

Whey Products and Lactose
  Processing Techniques
  Whey Products
  Lactose

Milkfat Concentrates
  Cream
  Butter and Its Products

Gelation—The process of gel formation, in which globular proteins act as gelling agents and provide desirable texture by holding a large quantity of water. Caused by heat denaturation or by pH changes, salt addition, or enzyme action.

Fatty acids—A group of chemical compounds characterized by a chain made up of carbon and hydrogen atoms and having a carboxylic acid (COOH) group at the end of the molecule. When they exist unattached to other compounds, they are called free fatty acids.

available to remove water from milk. A significant reduction in volume saves handling, packaging, and transportation costs.

CONDENSED PRODUCTS

Condensed skim milk is commonly used as a source of milk solids in dairy applications and in the manufacture of ice cream, frozen yogurt, and other frozen desserts. Condensed whole milk is purchased largely by confectionery industries. Evaporated milk is a heat-sterilized product packaged in cans and finds limited uses because of its yellowish color and cooked flavor.

Condensed milk/skim milk. Raw milk is standardized to the desired ratio of milkfat to milk solids-not-fat (MSNF). A common concentration factor reduces the original volume of milk to about one-third, yielding 25–40% solids in the finished product. The concentration factor is selected so that, on cooling to 3.3–4.4°C

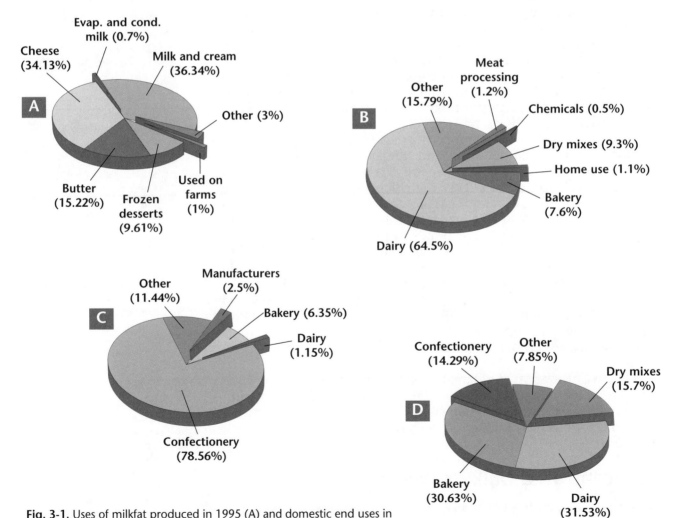

Fig. 3-1. Uses of milkfat produced in 1995 (A) and domestic end uses in 1995 of nonfat dry milk (B), dry whole milk (C), and dry buttermilk (D).

(38–40°F), the viscosity of the product poses no pumping or settling problem and lactose crystallization is minimized.

The milk is preheated to 93–96°C (199°F) for 10–20 min to destroy inherent enzymes and microbial load and to increase the heat stability of milk or control the viscosity of condensed milk. Next, the hot milk is drawn into a multi-effect evaporator, where it is concentrated using moderate temperatures and high vacuum. Milk boils under vacuum at 46–52°C (115–126°F). In the manufacture of condensed milk products, the boiling point rises by 0.5 degrees C (0.9 degrees F) for every doubling of the concentration.

The vapor is continuously removed until the desired concentration is achieved. The product can be homogenized, if desired, and is then cooled and packaged.

Condensed milks are generally customized orders. User plants specify total solids concentration, fat level, heat treatment, and processing conditions. Dairy concentrates offer economics of transportation costs and storage space. They must be transported and stored at 4.4°C (40°F) and used within five days to preserve quality. Bulk transportation of condensed milk to a food processing plant is a common practice.

Typical standard *specification* requirements for condensed skim milk are given in Appendix C, Table C-1.

Sweetened condensed milk/skim milk. Sweetened condensed milks contain 60% sugar in the water phase, which acts as a preservative, significantly enhancing keeping quality. Bacterial and mold growth is largely controlled in these products, but prolonged storage could cause spoilage by sugar-fermenting yeasts.

In the manufacture of sweetened condensed milk, milk is standardized for fat level, pasteurized, and homogenized. Sugar is then blended at 87.8°C (190°F) and evaporated under vacuum. Lactose reaches a supersaturated stage and crystallizes out of solution. If the crystals are too large, they impart an undesirable sandy texture. Crystal size is controlled by the addition of finely ground lactose during cooling. This step helps to create a very large number of extremely fine lactose crystals that cannot settle or be detected by taste. Sweetened condensed milk is used largely in confectionery manufacture.

DRY MILK PRODUCTS

Drying. Dry milk may be obtained from skim milk, partially skimmed milk, or whole milk (Fig. 3-2). In the first stage, milk is condensed to 45–50% solids under vacuum. (For dry whole milk, milk is homogenized before condensing.) The second stage involves drying in a spray dryer or on a roller dryer. Roller drying results in more scorched particles and poorer solubility of the powder than spray drying and is relatively rare. Spray drying gives dry milk of excellent solubility, flavor, and color. From the pan, condensed milk is pumped via an atomizer into a spray dryer at about 21°C (70°F). The operating conditions (e.g., preheating treatment and inlet and out-

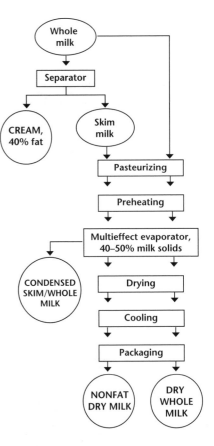

Fig. 3-2. Manufacturing outline for condensed and dry milk.

Specifications—A set of chemical or physical quality requirements that a product must meet before it is accepted.

TABLE 3-1. Typical Compositional Ranges (%) for Dry Milks

Constituent	Nonfat Dry Milk	Dry Whole Milk	Dry Buttermilk
Protein	34.0–37.0	24.5–27.0	30.0–33.0
Lactose	49.5–52.0	36.0–38.5	46.5–49.0
Fat	0.6–1.25	26.0–28.5	4.5–7.0
Ash	8.2–8.6	5.5–6.5	8.3–8.8
Moisture	3.0–4.0	2.0–4.5	3.0–4.0

let air temperatures) of the spray dryer are important in determining the functional quality of the product.

The atomizer may be of the pressure spray atomizing or the centrifugal type. The viscosity of the concentrated milk affects the shape, size, and consistency of the particles. Normal spray dryers yield droplets of milk, which, on drying, produce solid particles. (An alternative, foam drying, involves formation of bubbles of air or nitrogen and gives particles of higher surface area, which enhances solubility.) Variables such as the solids level in the milk concentrate, temperature, volume, and rate of air movement determine the speed of drying. Smaller drops of milk dry faster than the larger particles and result in a powder consisting of smaller particles. An atomizer producing uniform particle size gives a better-quality powder.

Coordination of atomizer output with dryer capacity is essential. If the dryer is incapable of removing adequate levels of moisture

Box 3-1. Grading Requirements

Grading requirements include the following (3): All NFDM, instant NFDM, dry whole milk, dry buttermilk, and dry buttermilk product for human consumption must conform to federal and state government regulations. Plant and processing equipment must be maintained in a strict sanitary condition. The product must be free from extraneous matter as described in the Federal Food, Drug and Cosmetic Act. It must be made from fresh, sweet milk to which no preservative, alkali, neutralizing agent, or other chemical has been added and which has been pasteurized. The dry milk product must be reasonably uniform in composition, white or cream colored, free from the brown or yellow color typical of overheated product or any other unnatural color, and substantially free of brown specks. Its flavor and odor in the dry form or on reliquefication must be sweet, clean, and free from rancid, tallowy, fishy, cheesy, soapy or other objectionable flavors and odors. The presumptive *coliform* estimate of the dry milk product must not exceed 90 colony-forming units (CFU)/g, except for instant NFDM and whole milk, in which it must not exceed 10 CFU/g. The product must be packed in containers that maintain quality with respect to sanitation, contamination, and moisture content under the customary conditions of handling, transportation, and storage.

The *phosphatase test*, when run at the option of the U.S. Department of Agriculture for official grading purposes or when requested by the buyer or seller, must show a test reading of not more than 4 µg of phenol per milliliter of reconstituted product.

Most user plants specify that dry products contain no detectable *Listeria*; *Salmonella*; coagulase-positive *Staphylococcus*; or antibiotic, pesticide, or herbicide residues. When the dry product is used in whipped products like ice cream or frozen yogurt, a specification may be added to ensure the absence of antifoaming compounds, which could interfere with getting the required level of air incorporated into a whipped food product.

from milk particles, the powder contains excessive moisture. On the other hand, if the atomizer supplies insufficient milk to the dryer, overheated powder results in poor solubility. In the case of whole milk powder, such conditions would result in fat separation and losses during handling and packaging.

Moderately high air velocity in the drying chamber is desirable. Air volume and temperature also influence the final composition and functional properties of the powder. Air is generally heated with natural gas to obtain an inlet temperature in the range of 148.9–232.2°C (300–450°F).

In a box dryer, most of the powder accumulates on the floor of the dryer. However, a significant quantity of the powder remains suspended in the air. Powder is recovered from the air by filters and *cyclones*. Energy conservation processes recover heat from the air before its discharge into the atmosphere.

The hot powder is removed from the drying chamber and quickly cooled. Failure to cool promptly results in defects such as lumping, whey protein denaturation, discoloration, and a scorched flavor.

Typical compositional ranges for dry milks are given in Table 3-1. Grading requirements are described in Box 3-1.

Nonfat dry milk. NFDM is defined (2,3) as the product resulting from the removal of fat and water from milk and containing the lactose, milk proteins, and milk minerals in the same relative proportions as in the fresh milk from which it was made. It contains not over 5% by weight of moisture and not over 1.5% by weight of fat unless otherwise indicated.

Appendix C, Table C-2 shows specific grade requirements for both spray-dried and roller-dried NFDM.

Spray-dried NFDM comes in extra-grade or standard grade. Extra-grade denotes the highest quality. It is entirely lump-free. The reconstituted product may have a slightly chalky, cooked, feed, or flat flavor. The roller process product may also have a slight scorched flavor. Standard grade is also lump free. The reconstituted product may have slight bitter, oxidized, stale, storage, utensil, or scorched flavor. In addition, it may have a definite degree of chalky, cooked, feed, or flat flavor.

The selection of NFDM for various food products is determined by its functionality, which is related to the heat treatment received during its manufacture. For example, bakery applications require high-heat NFDM to prevent loaf volume depression. High heat imparts a high moisture-absorbing quality to the ingredient, which is desirable in meat, confectionery, and bakery products. On the other hand, a low-heat product possesses optimum sensory characteristics and is ideal for use in dairy products and beverages. Medium-heat powder is used in ice cream and other products in which water absorption and flavor are important. The *whey protein nitrogen test* is used as a criterion for heat classification. This test is based on measurement of whey proteins left intact following the heat treatment during the manufacture of dry milk. The high-heat product contains not more

Cyclone—A centrifugal device for separating materials (e.g., particles from air).

Coliform count—A group of intestinal tract microorganisms that, if present in food or water, usually indicates the contamination of that food or water with fecal matter.

Phosphatase test—Commonly used test for confirming properly pasteurized milk and milk products. It measures residual phosphatase, which would have been inactivated by proper heat treatment.

Whey protein nitrogen test—Test used as a measure of the degree of heat received during processing of dry milk and whey protein concentrates. It is correlated with protein denaturation and with certain functional characteristics of dairy concentrates.

Box 3-2. Titratable Acidity

Titrable acidity is used as a measure of quality in dairy products. It is composed of "apparent" and "developed" acidities. Fresh milk should have no significant amount of lactic acid present, since lactose should not have been decomposed by bacterial growth or severe heat treatment. However, when fresh milk is titrated with standard alkali, a certain volume of the base is needed to achieve the end point of titration as indicated by change of color of phenolphthalein (at pH 8.6). This "apparent acidity" is the titration of the acidic components other than lactic acid. The approximate contributions of various milk constituents to the apparent percent titratable acidity (%TA) of milk (expressed as lactic acid) are: carbon dioxide, 0–0.01; caseins, 0.05–0.08; whey protein, 0.01–0.02; phosphate, 0.06; and citrate, 0.01. Thus, an apparent titratable acidity of 0.13–0.18% is contributed by constituents other than lactic acid.

Developed acidity is the portion of the titratable acidity that develops as a result of bacterial production of lactic acid from the metabolic breakdown of lactose in an anaerobic condition. To determine titratable acidity, 9 g of a milk or its product is titrated with $0.1N$ sodium hydroxide to pH 8.6, the phenolphthalein end point, and %TA is expressed as lactic acid.

$$\%TA = \frac{ml\ of\ 0.1N\ alkali}{10}$$

Since titratable acidity is exclusively attributable to the constituents of serum solids or milk solids-not-fat, it is clear that high-fat products like cream and butter will display a lower %TA than milk. To calculate %TA of cream containing 40% fat, the following relationship is used:

$$\%TA\ of\ cream = \%TA\ of\ milk \times \frac{\%\ serum\ solids\ in\ cream}{\%\ serum\ solids\ in\ milk}$$

Example: If 4%-fat milk has a %TA of 0.16%, cream with 40% fat will have %TA as follows:

$$\%TA\ of\ cream = 0.16 \times \frac{(100-40)}{(100-4)} = 0.16 \times \frac{60}{96} = 0.10$$

The serum solids of cream can be calculated if the fat content of the cream is known. The solids content of the serum phase of various fluid dairy products is approximately 8.8%. Thus, 100 lb of 40%-fat cream will contribute 40 lb of milkfat and 5.28 lb of serum solids, as shown in the following calculation:

$$\%\ serum\ solids\ of\ 40\%\ fat\ cream = (100 - 40) \times 0.088 = 5.28\%$$

than 1.5 mg of whey protein nitrogen per gram; the medium-heat product contains 1.51–5.99 mg/g and the low-heat product not less than 6.0 mg/g.

For ease of dispersibility of milk in water, an instantization or agglomeration process is available. Instant NFDM is free-flowing and lump-free and reconstitutes readily in cold water. The process for its production involves incorporation of a small amount of moisture in dry milk particles suspended in air, forming porous aggregates. These agglomerated particles are then redried and ground. This treatment enhances the surface area, which facilitates its reconstitution. On reconstitution, it yields a liquid product with a sweet pleasant flavor. Slight chalky, feed, cooked, and flat flavors may be present. Appendix C, Table C-2 gives specifications for extra-grade instant NFDM.

Dry whole milk. Dry whole milk is defined (3) as the product resulting from the removal of water from milk and containing not less than 26% nor more than 40% milkfat and not more than 5.0% moisture (as determined by weight of moisture on a MSNF basis).

Reconstituted extra-grade whole milk powder possesses a sweet, pleasant flavor. It may have a slight degree of feed flavor or a definite degree of cooked flavor, but no off-flavors. The products should be free of graininess on reconstitution and exhibit no burnt particles.

Reconstituted standard-grade whole milk powder is also sweet and pleasant but may exhibit slight bitter, oxidized, scorched, stale, and storage-related flavors. Feed and cooked flavors may be definitely apparent. It is reasonably free of burnt particles and graininess.

Since the product contains milkfat at high levels, deterioration caused by *oxidation* can be prevented by packaging in nitrogen or carbon dioxide. The oxygen content in the package should be less than 3%.

Specifications for whole dry milk are given in Appendix C, Table C-3. Some optional standards include a copper content of 1.5 ppm maximum, an iron content of 10 ppm maximum, and *titratable acidity* of 0.15% maximum. If fortified, it contains 2,000 IU of vitamin A and 400 IU of vitamin D (per quart basis in reconstituted liquid). See Box 3-2 for an explanation of titratable acidity.

Casein, caseinates, and milk protein concentrates. Casein (Appendix C, Table C-4) represents products obtained from pasteurized skim milk by precipitation of the casein fraction of milk protein using an acid or the enzyme *chymosin*, followed by drying. Caseinates are derived from casein by treatment with a suitable alkali. Casein is basically insoluble in water, whereas caseinates are easily dispersible. Acid casein is produced by precipitation of skim milk with hydrochloric, sulfuric, acetic, or lactic acid at pH 4.6. Acid-precipitated casein is neutralized to pH 6.7 with sodium hydroxide for the production of sodium caseinate. Similarly, potassium hydroxide and calcium hydroxide yield potassium and calcium caseinates, respectively.

Milk protein concentrate is obtained by ultrafiltration of skim milk and subsequent spray drying. The protein content can vary to meet the requirements of the particular process cheese product.

Casein and caseinates furnish emulsification, whipping, and texture-modifying attributes. Sodium caseinate aids in the formation of a stable emulsion by locating itself at the interface of oil and water. Since casein has both water-soluble (*hydrophilic*) and fat-soluble (*lipophilic*) portions in the molecule, fat globules in the emulsion get a coating of caseinate, which provides stability to the emulsion.

Whipping ability is related to the ability of caseinate to migrate to the air-liquid interface. Casein coats the air bubble, giving a fairly stable foam. Foam stability is enhanced by the presence of whey proteins or *hydrocolloids* in the food system.

The texture-modifying property of caseins is dependent on water-binding capacity, which, in turn, leads to increased viscosity. Sodium caseinate helps in preventing *syneresis* or water separation during freeze-thaw cycles or autoclaving of stabilized emulsions. Calcium caseinate strengthens the protein matrix of the food system, simulating cheese structure. Uses of caseins include dietetic food, infant formula, desserts, dressings, soups, sauces, coffee whiteners, cream liqueur, caramels, toffees, breads, cookies, meats, pasta, cereals, frozen dairy products, process cheese, and imitation cheese.

Oxidation—A chemical reaction in which the double bond on a lipid molecule reacts with oxygen to produce a variety of chemical products. The consequences of this reaction are loss of nutritional value and formation of the off-flavors associated with rancidity.

Titratable acidity—Test used for determining milk quality and for monitoring the progress of fermentation in cheese and fermented milks. It measures the amount of alkali required to neutralize the components of a given quantity of milk and milk products and is expressed as percent lactic acid.

Chymosin—A milk-clotting enzyme found in rennet and now derived by biotechnology.

Hydrophilic—Attracted to water (polar).

Lipophilic—Attracted to fat (nonpolar).

Hydrocolloids—Gums; water-soluble or modified polysaccharides used for thickening and water binding.

Syneresis—The separation of liquid from a gel; weeping.

Hygroscopic—Readily taking up and retaining moisture.

Sweet whey—Water and milk solids left after removal of curd in the manufacture of Cheddar, Swiss, and mozzarella cheeses. Its pH is about 5.5–6.0.

Acid whey—Water and milk solids left after removal of curd in the manufacture of cottage and ricotta cheeses. Its pH is about 4.4–4.6.

Whey Products and Lactose

Whey, the greenish-yellow liquid produced from the manufacture of cheese, contains about half the solids of whole milk. Its composition depends largely on the variety of cheese being made. The solids are valuable additions to the functional properties of various foods, as well as a source of valuable nutrients.

In the manufacture of nonhygroscopic dry whey products (4), the lactose in condensed whey is allowed to crystallize before drying. The majority of the lactose crystallizes to the (-monohydrate form. If lactose is present in the amorphous state, the resulting powder is *hygroscopic*. The objective is to reduce or eliminate the undesirable amorphous form of lactose in the powder. The powder obtained by inducing crystal formation in condensed whey is nonsticking and stable.

PROCESSING TECHNIQUES

Techniques for solids recovery. Concentration reduces the amount of water, thereby lowering shipping costs through reduced bulk. It also improves keeping quality and provides a product more suitable for direct use in foods. The cost of removing a pound of water in an efficient evaporator may be about one-tenth the cost of removing it in a spray dryer. This consideration has encouraged the development of more uses of whey and whey fractions in concentrated form. One major development has been the concentration of whey or whey fractions to 65–70% solids. This causes sufficient lactose crystallization to tie up the rest of the moisture, causing solidification into preformed blocks for use as animal "lick blocks."

Drying gives maximum concentration, extends storage stability, and provides a product amenable to incorporation into food. With a proper dryer, dairy processors convert *sweet whey* into a stable, non-hygroscopic, and noncaking product. In the first stage of this process, high-solids whey concentrate is spray dried to a free moisture content of 12–14%, causing lactose to take on a molecule of water and become crystallized. This causes whey solids to convert from a sticky, syrupy material into a damp powder with good flow characteristics. In the second stage, the powder is dried to approximately 4% moisture, as described below for acid whey drying. For drying acid (cottage cheese) whey, a commercial dryer combines spray drying with through-flow continuous-bed drying. The concentrate is spray dried in the hot air chamber to 12–15% moisture. The particles fall to a continuous, porous, stainless-steel belt, where lactose undergoes rapid crystallization. Crystallization of lactose before final drying is mandatory for drying *acid whey*. A belt conveys the product to another chamber, where the whey is further dried by dehumidified air that moves through the porous bed.

Fractionation techniques. Membrane technology (including ultra-filtration [UF], reverse osmosis [RO], and electrodialysis) and *ion-*

exchange techniques have resulted in the development of highly functional ingredients.

Excessive mineral content makes whey distasteful, and minerals can have an adverse effect on the physical properties of some foods. The two most widely used demineralization processes for whey are ion exchange and electrodialysis.

Ion exchange. In the ion-exchange process, whey is passed through two containers filled with special synthetic resins that have the ability to exchange ions. In the first container, the resins exchange hydrogen ions for *cations* in the whey. Here the positive ions of the salt are captured, and acid is formed by the release of hydrogen ions. The whey then goes to the second container and is passed over the anion exchanger, where hydroxyl ions are exchanged for negative ions of the salt, and water is formed.

When the mobile ions of the resins are completely replaced by other ions, the resin must be regenerated for further use. This is done by passing an acid (hydrochloric) solution through the cationic exchanger and a basic solution (NaOH) through the anionic exchanger.

Electrodialysis. This technique, a combination of electrolysis and dialysis, is the separation of *electrolytes*, under the influence of an electric potential through semipermeable membranes. The driving force is an electric field between the anode (positively charged) and the cathode (negatively charged). Several ion-selective membranes, each of which is permeable only to anions or to cations, are placed between the anode and the cathode. Every second membrane has a positive charge, repelling positive ions and allowing negative ions to pass. In between are negatively charged membranes doing just the opposite.

In principle, whey is pumped through every second space between two membranes, and a solution of NaCl (cleaning solution) is pumped through the compartments between the whey streams. The ions move from the whey stream into the cleaning solution, where they are retained because they cannot move any further. The cleaning solution contains minerals, acids, some lactose, and small nitrogenous molecules. The membranes are cleaned chemically. Protein molecules remain in the fluid while the minerals are removed. The process results in a protein concentrate.

Reverse osmosis/ultrafiltration. Related RO/UF membrane processes have become major factors in the field of whey concentration and fractionation. The two are pressure-activated processes that separate components on the basis of molecular size and shape.

RO is a process in which virtually all species except water are rejected by the membrane. The osmotic pressure of the feed stream in such a system will often be quite high. Consequently, to achieve adequate water flux rates through the membrane, such systems often use hydrostatic operating pressures of 5,883.6 kg/cm^2 (600 psi) or greater. UF refers to the process in which the membrane is permeable to relatively low molecular weight solutes and solvent (permeate) but impermeable to higher molecular weight materials (retentate).

Ion—An atom or group of atoms that carries a positive or negative electric charge.

Cation—A positively charged ion.

Electrolytes—Positively (cation) and negatively (anion) charged ions.

TABLE 3-2. Standard Specifications for Dry Sweet and Acid Whey

Characteristic	Dry Sweet Whey	Dry Acid Whey
Moisture %	3.5–5.0	3.5–5.0
Protein, %	11.0–14.5	11.0–13.5
Fat, %	1.0–1.5	0.5–1.5
Lactose, %	63.0–75.0	61.0–70.0
Ash, %	9.8–12.3	9.8–12.3
Scorched particles, mg	7.5–15.0	7.5–15.0
Titratable acidity, %	0.10–0.15	0.35–0.44
Color	Off-white to cream	Off-white to cream
Flavor	Normal sweet whey	Slightly acid, whey
Standard plate count, CFU[a]/g	50,000 max.	50,000
Coliform count, CFU/g	10 max.	
Salmonella	Negative	Negative
Listeria	Negative	Negative
Coagulase-positive *Staphylococcus*	Negative	Negative

[a] Colony-forming units.

The permeability and selectivity characteristics of these membranes can be controlled during fabrication so that they retain only molecules above a certain molecular weight. Thus, while UF is essentially a fractionating process, RO is effectively a concentrating process.

One advantage of UF over other processes is that, by varying the amounts of permeate removed, a wide variety of protein concentrates, ranging up to 60% protein, can be obtained. Higher levels can be obtained by simultaneously adding fresh water and concentrating by UF. The permeate is used for manufacture of milk sugar (lactose) by condensing and crystallization. Lactose crystals are harvested and dried in a tumble dryer.

WHEY PRODUCTS

Dry sweet whey. Dry sweet whey (Table 3-2) is produced by drying defatted fresh whey obtained from Cheddar, mozzarella, and Swiss cheese manufacture. It contains all the constituents except water in the same relative proportion as in liquid whey.

This ingredient is widely used in bakery products, dry mixes, process cheese foods and spreads, frozen desserts, sauces, meat emulsions, confections, soups, gravies, snack foods, and beverages.

Dry acid whey. This is similar to dry sweet whey but is produced by drying fresh whey obtained from cottage and ricotta cheese manufacture. Dry acid whey has an additional functional attribute of providing acid flavor in frozen desserts, and it imparts desirable textural properties to bakery items.

Reduced-lactose whey. Reduced-lactose whey (Appendix C, Table C-5) is produced from whey by crystallizing a majority of the lactose out and recovering the mother liquor. The lactose content of the dry product is 60% or less. The product is used in the formulation of confections, prepared dry mixes, bakery products, soups, sauces, gravies, dry seasoning blends, infant foods, and meat and cheese products.

Reduced-minerals whey. This is produced from whey by selective removal of a portion of the minerals. The ash content of the dry product is 7% or less. It is useful in the same products as reduced-lactose whey, in which mineral profile and concentration are critical attributes.

Whey protein concentrates. Whey protein concentrates (Table 3-3) are products derived from whey by removal of minerals and lactose. The process of protein concentration utilizes UF, electrodialysis, and ion-exchange technologies. On a dry basis, the protein concentrate contains a minimum of 25% protein. Whey protein isolate contains at least 92% protein.

A whey protein concentrate of 34% protein is commonly used as a stabilizer and *fat mimetic* in yogurt, bakery mixes, dietetic foods, infant foods, and confections. Its water-binding ability, fatlike mouthfeel, and gelation property are particularly useful in these products. A whey protein concentrate of 50 or 80% protein is especially suited for use in nutritional drinks, soups, bakery products, meat, dietary foods, and protein-fortified beverages. It gives clear suspensions over a wide pH range and has a bland flavor. Some applications require undenatured ingredients to maximize water-binding capacity during food processing. It is also available in a gel-forming version. It can be used as an economical egg-white replacement in food formulation.

> **Fat mimetic**—A fat-replacing ingredient based on protein, starch, other carbohydrates, or hydrocolloids that mimics the properties of fat.

LACTOSE

Lactose (Appendix C, Table C-5) is crystallized from condensed whey or from the permeate (50–60% solids) obtained by UF fractionation of milk or whey. The supersaturated solution is cooled under specific conditions to crystallize lactose. The crystals are harvested and washed to remove the mother liquor and dried. Crude lactose obtained this way contains about 98% lactose. Edible and USP grades are produced from crude lactose by protein precipitation, decolorization with activated carbon, and subsequent demineralization. Lactose is further refined by recrystallization or by spray drying.

Crystalline α-lactose hydrate is hard and not very soluble. β-Lactose crystals are sweeter and more soluble. However, in solution, the two forms equilibrate to a 62:38 ratio in favor of α-lactose. Lactose in amorphous form is too hygroscopic. It possesses a mildly sweet taste. It is available as Crude, Edible/Food Grade, or Refined/USP Grade.

The grind size of lactose is important depending on its application. Typically, grind 200 is approximately

TABLE 3-3. Specifications for Whey Protein Concentrates

Standard	Whey Protein Concentrate	Whey Protein Isolate
Moisture, %	3–4	4.5
Protein, %	34–80	92
Lactose, %	10–55	0.5
Fat, %	1–10	1.0
Ash, %	4–8	2.0
pH	6–7	6–7
Scorched particles, mg	7.5–15.0	7.5–15.0
Color	White to cream	Cream
Flavor	Bland, clean	Bland, clean
Standard plate count, CFU[a]/g	<50,000	<50,000
Coliform, CFU/g	<10	<10
Salmonella	Negative	Negative
Listeria	Negative	Negative
Coagulase-positive *Staphylococcus*	Negative	Negative

[a] Colony-forming units.

TABLE 3-4. Typical Standards for Fluid Cream Products

Standard	Frozen Cream	Heavy Cream	Medium Cream	Light Cream	Half & Half	Plastic Cream
Fat, % min.	50.0	36.0	30.0	18.0	10.0	80.0
Solids-not-fat, %	4.3	5.5	6.1	7.1	9.6	1.7
Titratable acidity, %	0.08	0.10	0.11	0.13	0.14	0.04
Standard plate count, CFU[a]/ml	All have <5,000.					
Coliform count, CFU/3 ml	All have <1.					
Psychrotrophic count, CFU/10 ml	All have <1.					
Weight, lb/gal	8.31	8.37	8.436	8.524	8.37	...

[a] Colony-forming units.

Triacylglycerols—Compounds of glycerol with three fatty acids attached. Neutral fats are composed of mono-, di-, and triacylglycerols (triglycerides).

Lactones—Chemical compounds derived from the hydrolysis of hydroxy fatty acids. Constituents of the overall flavor of dairy products.

Methyl ketones—Constituents of the flavor profile of dairy products, similar to lactones. Derived from β-keto acids.

74 μm size; 95% of the product passes through a No. 30 sieve and 50% passes through a No. 60 sieve.

Applications include infant formula, health and geriatric foods, dietetic formulations, dry mixes, confections, drinks, candy, fermentation substrate, and dairy beverages. Lactose as a sugar is characterized by low solubility and low sweetness. It imparts a desirable brown color to bakery items and enhances flavor in beverages. It is a useful carrier of intense sweeteners and pharmaceutical preparations.

A product called *lactose product* is similar to Food Grade lactose except that it contains only 93–97% lactose and has higher levels of nonlactose dairy constituents.

Milkfat Concentrates

Separating whole milk into skim milk and cream leads to several dairy intermediate ingredients that furnish sources of milkfat or butterfat in various foods (2). *Triacylglycerols* constitute 95–96% of total milk lipids. Approximately 85% of total fatty acids consists of lauric, myristic, palmitic, stearic, and oleic acids. Although present in minor amounts, butyric, capric, caproic, and caprylic acids are main contributors to the flavor. Since the melting point of butterfat ranges between 30 and 41.1°C (86 and 106°F), 95% of milkfat melts almost completely at body temperature, delivering a clean mouthfeel without a waxy sensation. The phospholipid level is 0.8–1.0%. Phospholipids, which are concentrated in low-fat products and in buttermilk, are excellent emulsifiers.

Milkfat provides a unique aroma and flavor in ice cream, pies, cakes, and cookies. More than 120 flavor compounds have been identified in butter. *Lactones*, *methyl ketones*, aldehydes, and dimethyl sulfides contribute significantly to the flavor profile of butter. In addition, butterfat contains precursors that generate aroma and flavor compounds on heating, as observed during baking. Major flavor compounds include short-chain fatty acids, which are characteristic of milkfat. *Diacetyl* is considered a key flavor compound providing the rich note. Volatile short-chain fatty acids released during baking contribute to the aroma of baked goods and confections.

In addition to aroma and creamy flavor, milkfat has a characteristic mouthfeel and a unique melting profile. It provides *lubricity*, moistness, and a cooling sensation in the mouth. Furthermore, the physical properties of milkfat and cream provide structure in chocolates and perceived richness in soups and sauces; prevent *blooming* in chocolates, cookies, and pastries; and contribute to aeration in cakes and frostings and tenderness of flake in bakery products.

The choice of a milkfat concentrate depends on its role in the food product. Cream products furnish an emulsion of fat globules dispersed in a liquid phase, whereas butter gives a continuous-phase fat in which water droplets are dispersed.

Diacetyl—A chemical compound characterizing the flavor of butter, milkfat, and certain fermented dairy products.

Lubricity—A desirable slippery sensation in the mouth imparted by fats.

Bloom—A dusty white appearance on the surface of chocolate caused by the formation of certain types of fat crystals.

CREAM

Cream is prepared from milk by centrifugal separation. U.S. standards require cream containing a minimum of 36% fat to be labeled "heavy whipping" cream. Cream used as an ingredient contains 36–40% fat. By standardizing with skim milk, cream of different fat levels can be produced. Light whipping cream and light ("coffee" or "table") cream contain 30–36% and 18–30% fat, respectively. Specific homogenization and heat treatments bring about desirable grades of viscosity in cream products. Cream should be stored under refrigeration. It can be quick-frozen and stored frozen until used.

Table 3-4 gives standards and specifications for fluid cream products. Spray-dried cream, made from sweet cream, provides a creamy flavor in dough mixes. Appendix C, Table C-6 shows its specifications.

BUTTER AND ITS PRODUCTS

Butter, a concentrated butterfat product, is an important ingredient as such or may be converted to more stable ingredients such as butteroil or anhydrous milkfat (5). Figure 3-3 shows a flow sheet for manufacture of butter and its products, and Appendix C, Table C-7 gives standards for butter and spray-dried butter. Butter contains a minimum of 80% fat and approximately 17% water, 1.6% salt, and 1% MSNF. The MSNF is involved in cooked flavor notes when foods are cooked in butter, as is observed in brown melt sauces.

Processing. Butter is usually churned from cream at a temperature conducive to an optimum ratio of crystalline fat to solid fat to liquid fat. Batch-process churns use cream of 35–45% fat, while continuous-process churns require cream of 42–44% fat. The cream is pasteurized at 73.8°C (165°F) for 30 min or 86°C (185°F) for 15 sec. The pasteurized cream is cooled to ≈7.2°C (≈45°F) so that, at the end of fat crystalization (≈16 hr), it is 10°C (50°F) and ready for transferring to a clean and sanitized butter churn. Annatto coloring is added if desired.

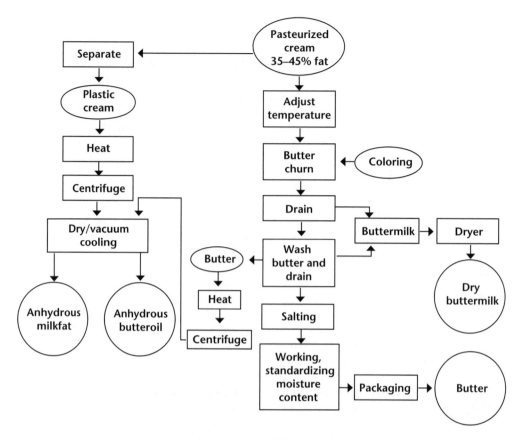

Fig. 3-3. Manufacturing outline for butter and butter products.

Phase inversion—The reversal of emulsion phases, changing the emulsion type, e.g., from the oil-in-water type to the water-in-oil type.

In a batch churn, the churn is rotated until butter granules the size of popcorn or peas are formed. Cream, an oil-in-water type of emulsion, undergoes *phase inversion* during churning to become a water-in-oil system. During churn rotation, cream foams and fat globules rupture to yield free fat, which cements the remaining globules, forming clumps. At this stage, the liquid phase, called buttermilk, is drained. Clean cold water (at a temperature 2–4 degrees F [1.1–2.2 degrees C] lower and at 25% of the volume of buttermilk drained) is added to the churn to rinse the butter. After draining, more water is added until the rinse is almost clear. The total volume of wash water equals the volume of buttermilk drained. The washed butter is mixed with enough salt to get 1.6% salt in the finished product.

The next step is called "working," the process of crushing the remaining fat globules and liberating additional free fat. The emulsion breaks and additional milkfat is freed, which becomes part of the continuous phase of the butter. Working involves churning until the body of butter is closely knit, assumes a waxy character, and shows no moisture pockets on the surface. The moisture content is adjusted to obtain a minimum of 80% fat in the butter. Working of butter continues until all the moisture is incorporated. Butter is then pumped and packaged.

Continuous butter churns accelerate the churning process. Washing of butter is not required. Cream of 42–44% fat content is churned in a cylinder, and buttermilk is drained. Butter granules are then worked and packaged. In another process, cream is concentrated by a separator to give 80% fat (*plastic cream*) before phase inversion; this is turned into butter by chilling and working.

Cultured cream butter made from cultured cream, has a distinct "butter" flavor as distinguished from sweet cream butter.

Light or reduced-fat butter. This product contains 40% fat and is equivalent in physical, sensory, functional, and nutritional characteristics to butter. If the light product differs significantly in performance characteristics from butter, the label must include a statement such as "not recommended for baking."

Spray-dried butter. Spray-dried butter is made in a process in which butter is melted, blended with nonfat dry milk, homogenized, and spray dried. Because this product is shelf stable, it may be an optional specialty ingredient in food plants manufacturing dry bakery, soup, and dessert mixes.

Plastic cream. Plastic cream contains 80% milk fat. It resembles butter in consistency but, in contrast to butter, is still an oil-in-water type of emulsion. It can be stored in frozen form and offers economy of transportation and storage. It can be used as an ingredient in cheese products and in frozen dessert formulation.

Buttermilk. Buttermilk is obtained as a by-product of butter production (Fig. 3-3). It represents the liquid phase of cream after churning. In addition to MSNF, it contains components of milk fat globule membranes liberated during the churning process. Accordingly, the phospholipid and total lipid content of dried buttermilk is significantly higher than that of NFDM. The phospholipids considerably reduce shelf life because they are prone to degradation, causing fishy odors and flavor defects.

The color of extra-grade buttermilk ranges from cream to light brown. It is free of lumps. The reconstituted product possesses a sweet, pleasing flavor and no off-flavors. Standard-grade buttermilk may have slight unnatural flavor but no offensive taints.

Dry buttermilk. Dry buttermilk is defined (3) as the product resulting from removal of water from the liquid buttermilk derived from the churning of butter. It is used as an economical substitute for NFDM in ice cream because of its enhanced emulsifying and whipping properties. It must contain not less than 4.5% milkfat and not more than 5% moisture and have a protein content of not less than 30%.

Specifications for spray-dried buttermilk are given in Appendix C, Table C-8. An optional test is called the *alkalinity of ash test*. All buttermilk grades display no greater than 125 ml of 0.1N HCl per

Plastic cream—An oil-in-water emulsion (cream) containing at least 80% milkfat.

Alkalinity of ash test—Test to determine whether neutralizers (Na_2CO_3, $NaHCO_3$, $NaOH$, etc.) were used to neutralize the cream used for manufacturing butter, buttermilk, and dry milk products. It involves titration of ash with standard acid.

Peroxide value—A number that indicates the level of per-oxides in a fat or oil; a measure of the amount of oxidation of the fat or oil.

Antioxidants—Compounds that inhibit lipid oxidation.

100 g. This test assures that neutralizers like carbonate, bicarbonate, or hydroxide were not used in the cream used in butter manufacture.

Another dry buttermilk available is called *dry buttermilk product*. This designation indicates that it does not meet the specification of a minimum protein content of 30%. Protein content must be shown on the label. Except for protein content, dry buttermilk product meets all other standards of dry buttermilk.

Anhydrous milkfat. Anhydrous milkfat is defined as the food product resulting from the removal of practically all moisture and MSNF from pasteurized cream or butter. Its flavor is bland; rancid, oxidized, or other objectionable flavors are absent. It contains at least 99.8% milkfat. Its moisture must be less than 0.1%, and not more than 0.1% of nonfat butter constituents (including salt) must be present. Salt itself cannot exceed 0.05%, and free fatty acids cannot exceed 0.3%. The *peroxide value* cannot exceed 0.1 meq/kg of fat. The copper content must be below 0.05 ppm. Specified *antioxidants* are permitted.

Anhydrous milkfat is made directly from plastic cream of 70–80% fat. By the application of heat, the emulsion is broken to liberate free fat. Separation of fat from the aqueous phase is done via centrifugation, which concentrates fat to 96–98%. Further heating and vacuum treatment remove the rest of the moisture.

Butteroil. Butteroil is defined as the food product resulting from the removal of practically all the moisture and MSNF from butter. It has a bland flavor and is free of rancid, oxidized, and other objectionable flavors. It contains not less than 99.6% milkfat, not more than 0.3% moisture, and not more than 0.1% other butter constituents, of which salt must not exceed 0.05% and free fatty acids must not exceed 0.5%. Its peroxide value cannot exceed 0.1 meq/kg. The copper content must be less than 0.05 ppm. Antioxidants and their maximum levels are specified by the FDA (7 CFR Part 58).

Anhydrous butteroil utilizes butter as a raw material. Melted butter is washed with water and centrifuged to attain 96–98% fat. Further processing is similar to that for anhydrous milkfat.

Cooked butter flavors are generated in the manufacture of ghee, a clarified butteroil produced by heat processing of butter or cream at 120°C (248°F) followed by centrifugal recovery of the milkfat.

Anhydrous milkfat and butteroil are manufactured from cream and butter, respectively, with no intense heat treatment. Consequently, they lack heated butter flavors. Enzyme-modified milkfat or lipolyzed butteroil, produced by lipase treatment, is a concentrated source of butter flavor for use in bakery, low-fat, and low-calorie caramel food products. Butteroil products are good carriers of herbs and spice flavors. The advantages of using anhydrous milkfat or butteroil are: 1) no splattering or burn-on during sautéing, 2) increased puff for pastries and pie shells, 3) aroma and flavor enhancement in bakery items, 4) control of fat bloom in chocolate candy, and 5) ease of melting, pumping, and mixing with other food ingredients.

These products are packaged in appropriate containers in which air is replaced with nitrogen or carbon dioxide. For extended shelf life, they should be held at 4.4°C (40°F).

References

1. International Dairy Foods Association. 1996. *Milk Facts. Cheese Facts. The Latest Scoop Worldwide.* The Association, Washington, DC.
2. DMI. 1995. *Dairy Ingredient Application Guide.* Dairy Management Inc., Rosemont, IL.
3. American Dairy Products Institute. 1990. *Standard for Grades of Dry Milks Including Methods of Analysis.* Bulletin 916, revised. The Institute, Chicago.
4. Carić, M. 1994. *Concentrated and Dried Dairy Products.* VCH Publishers, Inc., New York.
5. Hettinga, D. H. 1992. Butter and butter products. In: *Encyclopedia of Food Science and Technology*, Vol. 1. Y. H. Hui, Ed. John Wiley, New York.

Processing and Specifications of Dairy Foods

Cheese and Cheese Products

Cheese is a preserved food. The main milk components (proteins, fat, and minerals) are concentrated and protected from rapid deterioration by spoilage microorganisms. Cheese provides sound nutrition, variety, convenience of use, portability, food safety, and novelty in flavors and textures. With over 400 varieties available, cheese and cheese products are used as ingredients in entrees, side dishes, and ready-to-eat snacks. The market continues to grow at an annual rate of 4%. The typical composition of cheese is shown in Appendix A.

NATURAL CHEESES

Natural cheeses are made directly from milk (or whey) by coagulating or curdling milk, stirring and heating the curd, draining off the whey, and collecting or pressing the curd. The cheese is ripened, cured, or aged at a particular temperature (and humidity in some cases) for various periods to develop the flavor and texture characterizing the particular variety of cheese. Cheeses can be classified by moisture content (Table 4-1), which affects their texture and body, or by the process used for curing or ripening (Fig. 4-1).

The extensive use of cheeses in food products is attributable to their contributions of flavor and texture. Cheese for manufacture of food products may offer economical opportunities because it does not necessarily have to meet the specifications for the more expensive Retail Grade cheese. USDA grades for selected cheeses are shown in Table 4-2. Grading is based on cheese flavor, body and texture, color, finish, and appearance. Swiss cheese is graded additionally for its eyes and salt level. In grading of Colby cheese, color is not a factor.

Key composition parameters and uses of selected cheeses are shown in Table 4-3.

Processing. The production of Cheddar cheese is illustrated in Fig. 4-2. Raw milk is standardized to a casein-fat ratio of 0.7, pasteurized, and transferred to a cheese vat at 31.1°C (88°F). Color may be added

TABLE 4-1. Classification of Natural Cheese Based on Moisture

Very High Moisture (80–55%)	High Moisture (55–45%)	Medium Moisture (45–34%)	Low Moisture (34–13%)
Cottage	Mozzarella	Edam	Romano
Ricotta	Camembert	Brick	Parmesan
Impastata	Brie	Swiss	Dry ricotta
Neufchatel	Pizza	Cheddar	Gjetost
Cream	Blue	Provolone	Mysost
Tvorog/Quarg			

Rennet—An extract of the fourth stomach of a calf, kid, or lamb, used in the coagulation of milk during cheesemaking.

Galactose—A monosaccharide (simple sugar) liberated when lactose is hydrolyzed by the enzyme lactase.

as an optional ingredient. A 1% cheese culture is then mixed with milk. As the culture grows, acidity starts to build in the milk. After titratable acidity has risen by 0.05–0.1%, *rennet* (a coagulating enzyme) is added at the rate of 3 oz per 1,000 lb of milk. The milk sets to a firm gel. After about 20 min, the gel is cut by special knives into 1/4- or 3/8-in. cubes.

At this point, whey separates out from the curd. The acidity of the whey should be 0.1–0.12%. The next step is cooking, which involves raising the temperature of the vat contents from 31.1°C (88°F) to 36.7–38.9°C (98–102°F) until an acidity of 0.17–0.20% is obtained. The liquid portion (whey) is drained, and the solid portion (curd) is allowed to build higher acidity at 37.8–38.9°C (100–102°F). The whey fraction is separated to remove cream. It is then condensed and spray dried to produce dry sweet whey, which has emerged as an important ingredient in bakery items.

Curd starts to knit or mat as a slab, and the process of Cheddaring is terminated at an acidity of 0.45–0.6%. The matted slab is then milled to form small-sized cheese curd, salted, and pressed into blocks or barrels. The blocks are then packaged and ripened at 7.2°C (45°F) for a period varying from three months to one year (Table 4-4).

Relationship of characteristics to final properties. Functional properties are important for cheese applications. Because the chemical composition of the cheese determines its functionality, its composition is controlled by strict control of cheese pH, calcium content, fat content, moisture level, salt content, lactose/*galactose* content, and *proteolytic activity* during storage.

The importance of functional properties can be illustrated by considering mozzarella cheese. It is popular for pizza and Italian foods because pizza made with mozzarella displays desirable stretch, melt, browning, and surface blisters and does not show excessive *oiling-off*.

Stretchability depends on the calcium-phosphate level and the proteolysis. Too much calcium (predominantly dicalcium

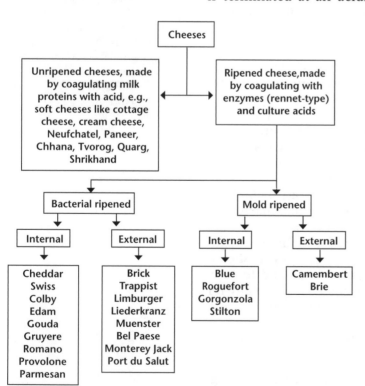

Fig. 4-1. Classification of cheeses by type of ripening.

paracaseinate) gives short stretch with a tough and grainy texture. Too much proteolysis in aged cheese also negatively influences stretchability.

Meltability is dependent on the proteolytic activity of the culture and on fat and moisture contents. Higher meltability is observed with higher fat and higher moisture content. Fresh mozzarella is relatively tough and elastic, but after ripening for about three weeks, the cheese develops desirable texture and meltability. Overmelting is also undesirable, because melted cheese runs off the pizza.

Browning, the result of Maillard reaction between a reducing sugar (lactose or galactose) and a free amino group of an *amino acid* in protein, is increased if too much galactose accumulates or too much lactose is retained or if proteolysis of amino acids is abnormal.

Excessive oiling-off is undesirable because it imparts an oily surface to the pizza. The fat content of the cheese and the strength of the protein matrix are critical factors in this regard. Pizza cheese is generally made from part-skim milk to reduce excessive fat, which might otherwise oil-off. Aged cheese tends to oil-off more than fresh cheese.

Blistering (size, number, and color) is a functional attribute of mozzarella cheese during baking. Large-sized blisters, caused by excessive ripening of the cheese, are undesirable. A young cheese generally has a large number of small blisters.

Shreddability is a function of fat and moisture content. Generally, low-moisture part-skim mozzarella has satisfactory shredding properties.

PROCESS CHEESE PRODUCTS

Natural cheese constitutes the main ingredient for the manufacture of *process cheese* products (1,2). Process cheese delivers fairly uniform flavor and texture, which are uncontrollable variables in natural cheese. Its melting characteristics can be manipulated by use of specific melting (or emulsifying) salts. Normal variations in flavor, typical of natural cheeses, are minimized by blending mild and strong flavors (and different ages) of natural cheese. Process cheese contains higher moisture than natural cheese.

In the manufacture of process cheese, selected cheeses are macerated and transferred to a cooking vessel. Emulsifying salts (commonly citrates and phosphates, up to a level of 3%) are used to prevent separation of fat during heat processing and also result in desirable body in the product. Also, other ingredients like common salt, preservatives, cream, dry milk, and whey may be added as desired. The mixture is heated to 73.8–82.2°C (165–180°F) for 1–5 min with vigorous agitation. Scraped surface equipment is necessary to facilitate heat transfer. The mixture turns fluid, and a homogeneous mass

TABLE 4-2. USDA Grades of Cheeses

Type of Cheese	USDA Grade
Cheddar	AA, A, B, C
Bulk American cheese for manufacturing	Extra Grade, Standard Grade, Commercial Grade
Monterey Jack	AA, A, B
Swiss	A, B, C, D
Colby	AA, A, B

Proteolytic activity (proteolysis)—The enzymatic breakdown of the protein molecule to yield peptides (smaller fragments).

Oiling-off—Liberation of free fat from cheese (e.g., pizza cheese) as a result of exposure to high temperature.

Amino acid—An organic acid containing both an amino (NH_2) and an acidic (-COOH) group; the building blocks of proteins.

Process cheese—A cheese product made by blending various natural cheeses in the presence of heat and emulsifying salts.

TABLE 4-3. Composition and Uses of Selected Cheeses

Cheese Variety	Fat (minimum %, dry basis)	Moisture (maximum, %)	Main Uses
Cream	33[a]	55	Crackers, cheese balls, bagels, spread, cheesecake desserts, appetizers
Blue	50	46	Salads, dips, appetizers, dressings
Cheddar (mild to extra sharp)	50	39	Sandwiches, salads, sauces, pizza, vegetables, snacks, entrees
Colby	50	40	Snacks, entrees, sandwiches
Gouda	29	45	Dessert, cheese tray, salads, appetizers
Monterey Jack	50	44	Sandwiches, cheese trays, Mexican foods, sandwiches, flavored cheeses
Parmesan	32	32	Spaghetti, pizza, minestrone soup, salad, lasagna
Romano	38	34	Same as Parmesan
Provolone	45	45	Italian food, desserts, sandwiches, appetizers
Swiss	43	41	Sauces, fondues, appetizers, sandwiches, desserts, soups
Mozzarella	45	60 (52 min.)	Pizza and Italian food, snacks
Low-moisture	45	52 (45 min.)	
Part-skim	30 (45 max.)	60 (52 min.)	
Low-moisture part-skim	30 (45 max.)	52 (45 min.)	

[a] As-is basis.

is obtained. The product is ready for packaging into forms and cooling to get process cheese loaves. For cheese slices, molten cheese is cast on a roller drum, followed by cutting into ribbons and slices and packaging.

Pasteurized process cheese. This food is prepared by comminuting and mixing, with the aid of heat, one or more cheeses of the same or different varieties with an emulsifying agent, producing a plastic homogeneous mass. (Cheeses *not* used for this process are cream cheese, Neufchatel cheese, cottage cheese, creamed cottage cheese, cook cheese, hard grating cheese, semisoft part-skim cheese, part-skim spice cheese, and skim milk cheese for manufacturing.) Heating is at not less than 65.5°C (150°F) and for not less than 30 sec.

The moisture content of process cheese made from a single variety of cheese must not exceed 1% more than that of the constituent natural cheese and not exceed 43%. For instance, a process cheese made from 39%-moisture Colby could contain up to 40% moisture. Fat in the dry matter (FDM) is similar to that of the natural cheese: in general, not less than 47% (Cheddar: not less than 50%; Swiss: not less than 43%, Gruyere: not less than 45%) (1).

Process cheese made from two or more varieties of cheese has a moisture content not exceeding 1% more than the arithmetical average of the maximum moisture content of the constituent cheeses, but in no case more than 43%. However, process cheese made from various varieties of Cheddar cheese cannot exceed 40% and that made from Swiss and Gruyere varieties cannot exceed 44%. The fat content (FDM) in this case is also the average of the FDM levels of the constituent cheeses as long as it does not fall below 47%, except that the product made from Swiss and Gruyere cheese contains FDM not less than 45%.

The consistency of process cheese is semifirm, creamy, and smooth as compared to its natural cheese counterparts. Functional properties that can be created are sliceability, extra melt (the ability to melt easily on heating, not thicken, and withstand high temperature for long periods), and slow melt (the ability to maintain shape at high temperature). Process cheese may be flavored with seasonings.

Pasteurized process cheese food. This cheese is similar to pasteurized process cheese except that it must contain moisture not exceeding 44% and fat content of not less than 23% (as-is basis) (2). Fat in dry matter is 41% minimum. It contains optional dairy ingredients: cream, milk, skim milk, buttermilk, cheese whey solids, anhydrous milk fat, dehydrated cream, whey protein, or skim milk cheese for manufacturing. The pH is adjusted to not below 5.0 with vinegar, lactic acid, citric acid, phosphoric acid, or acetic acid. It cannot contain more than 3% emulsifying agents or more than 0.2% sorbic acid or 0.3% sodium/calcium propionate.

Pasteurized process cheese food is produced by blending American cheeses of different ages with nonfat dry milk (NFDM) and whey, followed by pasteurization. It melts quickly to give a smooth liquid. The cold product can be sliced easily. Major uses include entrees, au gratin potatoes, sandwiches, and Mexican dishes. It may be flavored with seasonings, smoke, pimento, jalapeno, salami, pepperoni, etc.

Pasteurized process cheese spread. Similar to process cheese food but spreadable at 21°C (70°F), it has a moisture content of 44–60% and a fat content of not less than 20% (as-is basis) (2). It may contain optional dairy ingredients, emulsifying agents, and gums, not to exceed 0.8%. Acids may be added to obtain a pH level not below 4.0. Sweetening agents may be used (sugar, dextrose, or corn sugars).

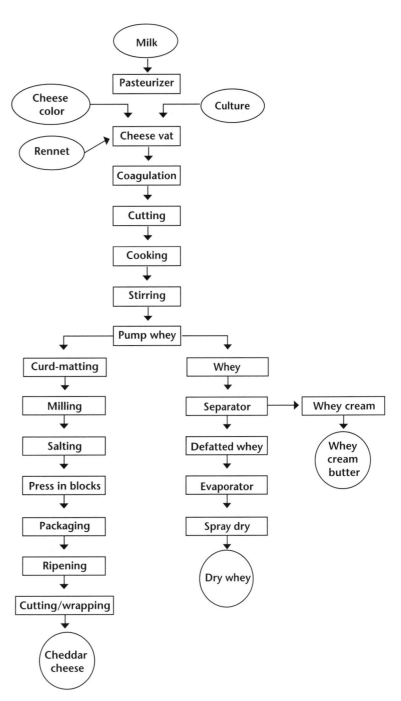

Fig. 4-2. Steps in cheese manufacturing.

TABLE 4-4. Typical Ripening Periods for Cheeses

Cheese Variety	Ripening Period
Blue (bleu)	2 months minimum
Brick	2 weeks
Cheddar	2 weeks to 12 months
Colby	2 weeks to 3 months
Cream cheese	Unripened
Gouda	3–6 months
Monterey Jack	2–6 months
Mozzarella	Unripened to 3 weeks
Muenster	2–8 weeks
Parmesan	10 months minimum
Provolone	2–12 months
Romano	5 months minimum
Swiss	2–9 months

Nisin—A metabolite of a cheese-fermenting micro-organism, which acts as a preservative.

Sorbic acid (<0.2%) or propionates (<0.3%) may be used as preservatives. *Nisin* (not more than 250 ppm) is permitted in the spread.

Process cheese spread may contain meat, vegetables, pimento, or pineapple or be flavored with blue cheese, onion, etc. Its uses include snacks, deviled eggs, noodle casserole, meat balls, hot vegetables, sandwiches, sauces, and dressings.

Cold pack cheese (club cheese). Manufacture involves blending various cheeses without heating. Only cheese from pasteurized milk may be used (2). Its moisture content is the same as that of the individual cheeses in the blend; the fat content in dry matter is not less than 47% in most cheese except Swiss (not less than 43%) and Gruyere (not less than 45%). Cold pack cheese may contain acids to standardize the pH to not below 4.5. Sorbic acid or propionates (<0.3%) may be used as a preservative.

It spreads easily, may be smoke flavored, and is used in appetizers, snacks, and desserts.

Cold pack cheese food. This product is prepared by comminuting and mixing (without heating) cheeses and other ingredients such as cream, milk, skim, buttermilk, whey solids, anhydrous milk fat, dehydrated cream, skim milk cheese, and whey protein. Acids may be added to bring the pH to not less than 4.5. Sweetening agents (sugar, corn syrup solids) may also be used. Sorbic acid or propionates (≤0.3%) may be used as preservatives. Guar gum or xanthan gum may be used (≤0.3%). Moisture content cannot exceed 44%, and fat content is not less than 23%. Besides cheese, it may contain corn sweeteners, dairy ingredients, fruits, vegetables, or meats. It may be smoke flavored. It is more spreadable than cold pack cheese. Its uses are the same as for cold pack cheese.

OTHER

Cheese powders. Spray-dried cheese powders (Table 4-5) are widely used as seasonings and flavorings. They are produced by macerating cheese, dispersing in water at a 35–40% solids concentration, adding emulsifying salts, homogenizing, and spray drying. Foam spray drying is considered to give a superior flavored product with larger particle size. In addition to cheese, dry milk, whey, vegetable oils, salt, enzyme-modified cheese concentrate, color, and seasonings may be incorporated.

Cheese powders can be packed in nitrogen atmosphere to give a storage life of a year or so. Hard Italian cheese (e.g., Parmesan) is dried after grating in tray or belt dryers in which dry hot air is circulated to reduce moisture to less than 6%. After cooling, the cheese is ground and packaged.

Enzyme-modified cheeses. Enzyme-modified cheeses (EMC) are cheese flavor concentrates obtained by treating raw cheese curd with specific lipases and proteases along with fermentation with a cheese culture. Flavor concentration 10- to 20-fold as high as that of ripened cheeses develops in one to three days. The cheese paste is then heat treated to stop the biochemical reaction and cooled. The EMC may be purchased as a paste or blended with whey and dried as a spray-dried powder. It offers significant savings as a substitute for aged cheese in cheese-flavored crackers and other bakery items. Also, it is an economical ingredient in certain process cheese products (2).

Potential uses of cheese powders are in au gratin potatoes, snack coatings, dips, dry mixes, salad dressings, sauces, soups, and crackers. In addition, seasonings like smoky, bacon, sour cream, onion, nacho, jalapeño, or Italian may be mixed with cheese powder to enhance their application.

TABLE 4-5. Product Specifications[a] for Some Cheese Powders

Type	Moisture (%, max.)	Fat (%)	Salt (%)	Ingredients
Blue cheese	4	46 ± 2	5 ± 1	Blue cheese, disodium phosphate, citrate
Blue cheese blend	4	36 ± 2	7 ± 1	Blue and Cheddar cheese, buttermilk, whey, vegetable oil, salt, disodium phosphate, lactic acid, vinegar, monosodium glutamate
Parmesan cheese blend	4	16 ± 2	3.25 ± 1	Parmesan cheese, whey, disodium phosphate, lactic acid
Romano cheese blend	4	19 ± 2	3.75 ± 1	Romano cheese, whey, disodium phosphate, lactic acid
Process cheese	4	46 ± 2	5.5 ± 1	American cheese, disodium phosphate, cream, salt, lactic acid, color
Cream cheese	4	65 ± 2	4 ± 1	Cream cheese, tocopherols
Cheddar cheese	4	47 ± 2	5.0 ± 1	Cheddar cheese, disodium phosphate, salt, lactic acid
Cheddar cheese blend	4	21 ± 2	9.5 ± 1	Cheddar cheese, whey, buttermilk, salt, color
Low-fat Cheddar cheese	5	8 ± 2	6.5 ± 1	Whey, buttermilk, maltodextrin, Cheddar, blue cheese, cultured nonfat dry milk, whey, salt, disodium phosphate, citric acid, color
Sauce mix	5	20 ± 2	8.0 ± 1	Cheese solids, whey, vegetable oil, maltodextrin, salt, buttermilk, sodium phosphate, nonfat dry milk, starch, citric acid, color

[a] In addition, microbial analysis must show a standard plate count of <75,000 colony-forming units (CFU)/g, yeast/mold count of <100 CFU/g, and coliform count of <50 CFU/g. *Salmonella* and coagulase-positive *Staphylococcus* must be negative.

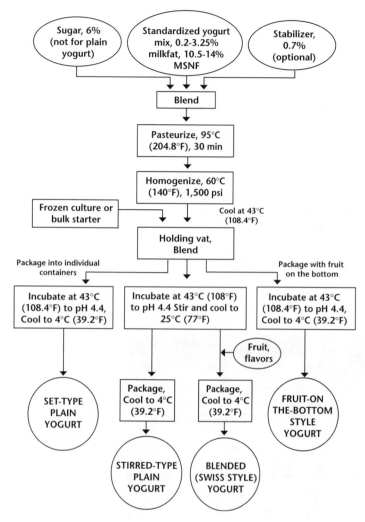

Fig. 4-3. Steps in yogurt manufacture. MSNF = milk solids-not-fat.

Cheese sauces. These are aseptically processed viscous liquids that are canned for convenient use as a dip or as a sauce on nachos, potatoes, etc. Typically, ingredients used are Cheddar cheese, skim milk, whey, buttermilk, vegetable oil, starch, sodium phosphate, salt, caseinate, citrate, color, lactic acid, stabilizers, emulsifier, and seasonings. They may contain 12% fat. The viscosity standard is 30,000–60,000 cps at 21.1–26.7°C (70–80°F). The shelf life of unopened containers is about a year.

Fermented Milks

Fermented milks, including yogurt, buttermilk, and sour cream, have long formed a vital segment of the human diet. The typical composition of these dairy ingredients is shown in Appendix A.

Fermentation processes extend the shelf life of milk nutrients and produce an array of flavors and textures. Major factors governing the flavor and texture development are composition of the dairy mix and nature of the culture employed. In addition, the cultures produce certain metabolites such as lactic acid, anticarcinogenic compounds, and a possible anticholesteremic factor, which render cultured milks apparently beneficial from a functional food standpoint. Furthermore, the microbial lactases of lactic cultures in cultured dairy foods aid in lactose digestion. The lactic cultures tend to prevent Gram-negative bacteria (responsible for many intestinal disturbances) from gaining predominance through competitive inhibition.

The manufacturing principles of yogurt are outlined in Figure 4-3. Figure 4-4 shows a fermenter for yogurt. Processing guidelines for different types of products are given in Appendix C.

Direct acidification procedures (avoiding the use of bacterial cultures) for buttermilk and sour cream have been developed.

Fermented milks are important ingredients in a wide variety of food products and popular cuisines. Yogurt, sour cream, and buttermilk are used in the formulation of salad dressings, sauces for pasta, and Mexican meals. Buttermilk gives pancakes and bakery items a fluffy and lighter texture. Cream cheese is an ingredient of cheesecakes.

Fig. 4-4. Fermentation tanks for yogurt manufacture. (Courtesy of APV)

Ice Cream and Frozen Desserts

Ice cream is a food produced by freezing a pasteurized mix while stirring. It is a frozen blend of air, water, milkfat, milk solids-not-fat (MSNF), sweeteners, stabilizers, emulsifiers, flavors, and coloring materials. Frozen dessert mixes are formulated, processed, and extruded through ice cream freezers to deliver consumer-desired attributes of flavor, texture, and shelf life.

FORMULATION OF ICE CREAM

The wide variety of ingredients used for production of different kinds of frozen desserts is defined by U.S. Food and Drug Administration regulations (2). Mellorine is similar to ice cream except that it contains no milkfat, which is replaced with vegetable fat in the formulation. Table 4-6 lists optional ice cream ingredients.

Frozen desserts may be labeled low-fat or nonfat depending on their fat contribution per serving of half a cup (65–70 g). Products containing no more than 3 g of fat per serving are classified as low-fat. Nonfat products contribute less than 0.5 g of fat per serving. Reduced-fat products provide less than 25% of the amount of fat in the reference product.

Dairy ingredients constitute 50–55% of the total solids of ice cream and related frozen desserts. The choice of dairy ingredients and the formulation of an ice cream mix are determined by regulatory standards, the desired quality of the frozen dessert, marketing strategy, consumer demand, relative prices, and availability of the ingredients in a given locality.

Milk products constitute the most important components because they furnish the basic ingredients for a good-quality ice cream. Variables related to dairy ingredients exert a profound influence on the flavor, body, and texture of the frozen product (see Box 4-1). The

TABLE 4-6. Optional Dairy and Related Ingredients for Ice Cream and Frozen Custard

Dairy Ingredients	Caseinates[a]
Cream	Casein precipitated with gums
Dried cream	Ammonium caseinate
Plastic cream/concentrated milkfat	Calcium caseinate
Butter	Potassium caseinate
Butteroil	Sodium caseinate
Milk	
Concentrated milk	
Evaporated milk	**Hydrolyzed Milk Proteins[b]**
Sweetened or superheated condensed milk	Hydrolyzed casein
Dried milk	Hydrolyzed whey proteins
Skim milk	
Concentrated, evaporated, or condensed skim milk	
Superheated condensed skim milk	
Sweetened condensed skim or part-skim milk	
Nonfat dry milk	
Sweet cream buttermilk[c]	
Condensed or dried sweet cream buttermilk[c]	
Concentrated delactosed skim milk	
Dry/concentrated modified skim milk[d]	
Whey[e]	
Reduced-lactose or reduced-minerals whey[e]	
Whey protein concentrate/isolate[e]	

[a] May be added to ice cream mix containing not less than 20% milk solids. Liquid or dry form of caseinates may be used. Must be free of excess alkali.

[b] May be used as stabilizers at a level not to exceed 3% by weight of ice cream mix containing not less than 20% total milk solids, provided that any whey and modified whey products used contribute, singly or in combination, not more than 25% by weight of the total nonfat milk solids content of the finished food. Any whey or modified whey products contained in the milk protein hydrolysates must fall within the 25% limitation.

[c] Adjust with water to total solids of 8.5% and titratable acidity of 0.17%.

[d] Treated with calcium hydroxide and disodium phosphate. pH value range, 8.0–8.3.

[e] Subject to the 25% limitation (see footnote b).

Frozen custard—An ice cream containing at least 1.4% egg yolk solids. Also known as French ice cream.

nature and intensity of overall ice cream flavor is a function of the flavor quality of the individual constituents and subsequent processing treatment of the ice cream mix. Flavor defects in the ingredients cannot be alleviated during ice cream making. Actually, flavor problems could be compounded as a consequence of negligent processing procedures.

The body or consistency of ice cream is related to the mechanical strength of the mix and its resistance to melting. Heat shock resistance is dependent on the nature and concentration of the stabilizer-emulsifier system used. The texture of ice cream depends upon the size, shape, number, and arrangement of air cells, fat globules, and ice crystals and the ratio of frozen and liquid water in the ice cream.

Balancing quality with cost is a major challenge to an ice cream manufacturer. Satisfactory composition produces an ice cream having an optimum combination of cost, flavor, body, texture, cooling effect, viscosity, whipping ability, and freezing characteristics. In summary, factors responsible for overall ice cream quality are: raw material quality, sanitary care during mix preparation, processing parameters, flavoring used, freezing techniques, and storage conditions.

Formulation of an ice cream mix involves utilization of both the fat and solids-not-fat components of milk. The functions and preferred sources of major ice cream ingredients are summarized in Table 4-7.

Sources of fat (3,4). *Sweet fresh cream and fresh milk.* Whole milk may be used primarily as a source of serum solids, but it will supply milkfat as well. Without question, there is no better source of fat than sweet cream because of its desirable flavor, convenience of handling, and good whipping characteristics. Fresh cream is judged by flavor, acidity, and bacterial count. The titratable acidity should be low and show no evidence of developed acidity. When fresh cream is not available at a favorable cost, alternative sources of fat should be considered.

> ## Manufacturing Procedure for Ice Cream Mixes
>
> Pasteurization should be at 80.5°C (175°F) for 25 sec. To homogenize, the equipment manufacturer's directions should be followed. Typically, the second-stage valve pressure is set at 500 psi and then the first-stage valve pressure is set to obtain a final gauge reading of 2,500 psi. The product is cooled immediately to 4.4°C (40°F) or below.

Frozen cream. The high price of sweet cream during certain seasons of the year makes storage of cream during the months of surplus economically attractive. All known precautions must be used to ensure prevention of the development of off-flavors in stored cream. Only the best cream should be processed for storage, and it should contain no developed acidity. Off-flavors likely to develop in frozen cream are rancid, fishy, oily, and tallowy. *Hydrolytic rancidity* is due to free butyric acid from the partial hydrolysis of butterfat brought about by enzymatic activity of lipase on the butterfat or by certain bacteria or the enzymes they produce. A proper heat treatment regime, an essential phase for the preparation of cream for freezing, consists of heating cream at 76.7°C (170°F) for 20 min, or 82.2°C (180°F) for 10 min, or 87.8°C (190°F) for 5 min. This treatment not only inactivates the lipase enzyme naturally present in milk but also destroys 95–99% of the bacteria present.

Heat treatment of cream also increases the resistance of the cream to oxidation. A fishy flavor in dairy products results from the formation of *trimethylamines* by the hydrolysis and oxidation of *lecithin*, a naturally occurring phospholipid in milk. Factors that promote development of this flavor are high acidity and the presence of *prooxidants* (iron or copper salts). Evidently, heat treatment at these times and temperatures "activates" or uncoils the proteins so that sulfhydryl groups are exposed and become oxidized by atmospheric oxygen in preference to the unsaturated fatty acids. Apparently, these sulfhydryl groups function as antioxidants in the liquid system. In addition they may complex with prooxidant minerals.

Following heat processing, the cream is quickly frozen. Proper packaging and handling of frozen cream are also important. Preferred packages include stainless steel or plastic containers. Quick-frozen cream is held at –23.3°C (–10°F) or lower. Disadvantages of frozen cream include the necessity of thawing before use and the fact that it is messy to handle.

Butter. Fresh, sweet-cream, unsalted butter can be used when appropriate. Often it is the cheapest source of fat. Sweet-cream butter has been found to whip more slowly because of the loss of natural emulsifiers (the phospholipids) in the buttermilk fraction. However, this is easily overcome with homogenization and com-

Hydrolytic rancidity—A flavor defect associated with the activity of the enzyme lipase on milkfat, which liberates short-chain fatty acids with objectionable odors.

Trimethylamines—Compounds responsible for a fishy odor in dairy products.

Lecithin—A phospholipid found in egg yolk and soybeans and also used as a food ingredient. It is a surfactant that can stabilize emulsions.

Fig. 4-5. An ice cream freezer. (Courtesy of APV)

Box 4-1. Overrun

An important variable related to foam formation as a result of aeration of ice cream mix during freezing is called *overrun*. This is the volume of ice cream obtained over and above the volume of mix used and can be calculated as follows:

Volume basis:

$$\% \text{ overrun} = 100 \times \frac{(\text{volume of ice cream made}) - (\text{volume of mix used})}{\text{volume of mix used}}$$

Weight basis:

$$\% \text{ overrun} = 100 \times \frac{(\text{weight of given vol. of mix}) - (\text{weight of equal vol. of ice cream})}{\text{weight of equal vol. of ice cream}}$$

Percent overrun is dependent upon the type of frozen dessert and the freezing equipment. Packaged ice cream may contain 70–75% overrun, while bulk ice cream may contain 90–100% overrun. Super-premium ice cream is generally in the 30–50% overrun range. Sherbet normally has 30–40%; milk shakes have only 10–50%. The official serving size of ice cream is one-half cup, and its weight varies from 60 to 70 g, depending on the overrun. Overrun is of much interest because it influences the texture as well as the cost of production of ice cream.

Prooxidants—Certain metal ions in dairy products that accelerate oxidative deterioration and the development of off-flavors.

mercially available emulsifiers. Butter can be stored frozen for an extended period.

Plastic cream. By re-separating sweet cream (fat content, 28–35%), in a special separator bowl at relatively high temperature (60°C, 140°F), a cream (oil/ water emulsion) is obtained with a fat content of 79–81%. Because of this high fat content, the product solidifies to a plastic or solid mass upon cooling and is therefore known as "plastic cream." If the product is cooled promptly to 4.4°C (40°F), it retains a liquid state. It can be packaged directly because crystallization of the fat is fairly slow and can be frozen if it is not to be used in the immediate future. Although plastic cream provides some savings from reduction in shipping weight and storage space, it is not widely used as a fat source.

Other. Butteroil and anhydrous milkfat are rarely used as fat sources in frozen desserts in the United States. Dry cream and dry whole milk are also not widely used in the formulation of ice cream mix because they are prone to flavor deterioration. During drying, the milkfat component of these products undergoes some degree of oxidation, leading to rapid deterioration during storage.

Sources of serum solids (3,4). *Fluid whole and skim milk.* Both are excellent sources and should be used in the mix whenever available at reasonable cost. However, because of their low serum solids content in contrast to the serum solids desired in an ice cream mix, their use is limited. Skim milk should be purchased on the basis of a definite MSNF content in order to guard against dilution with water.

Plain condensed skim milk. This source of serum solids is used more frequently than any other of the condensed products. Fresh condensed skim milk is easy and convenient to use, has an excellent flavor, and is easily available. The concentrate is paid for on the basis of the solids content, which runs around 25–35%. The heat treatment given fluid skim milk is usually the same as the regular pasteurizing range. The keeping quality of condensed skim milk is better than that of cream. It should be stored at 0–1.6°C (32–35°F) and used while fresh and sweet (usually for 7–10 days).

Plain condensed whole milk. This milk is concentrated about two and a half times and contains 8% fat and 20% serum solids. It is frequently used as an ingredient by ice cream manufacturers.

Superheated condensed skim or whole milk. The use of superheated condensed milk permits ice cream manufacture without stabilizers and may offer a marketing advantage with certain consumers. The already-condensed product is slowly heated to a high temperature, usually in the range of 82.2°C (180°F). When properly done, a concentrate of much greater viscosity is obtained, which improves the whipping ability of the ice cream mix and contributes a smooth texture. Superheating increases the hydration capacity of the milk proteins, which then bind more free water. Accordingly, less water is available to form ice crystals during freezing and shelf life, and the smooth texture of the ice cream is maintained throughout its shelf life. Superheating, therefore, functions like a stabilizer.

Sweetened condensed whole or skim milk. A sweetened condensed product is sometimes used as a source of MSNF. This ingredient provides 8.5% fat and 28% total milk solids. The added sugar (40–44%)

TABLE 4-7. Role and Sources of Various Components of Ice Cream

Component	Role and Function	Limitation	Sources in Order of Preference
Milkfat	Imparts desirable creamy rich flavor Source of fat-soluble vitamins Improves body texture Improves melting resistance	Calorifically dense Too much fat (17%) gives too much viscosity to mix and hinders whippability Source of oxidized, rancid, and fishy flavor defects	Fresh sweet cream Fresh milk Frozen cream Butter Plastic cream Butteroil Dry cream/milk
Milk solids-not-fat	Improves texture Imparts better body Permits higher overrun Source of protein, minerals, and vitamins	Relatively expensive component Improper levels cause "sandiness" defect Source of condensed milk flavor and cooked flavor defects	Fluid whole milk Fluid skim milk Condensed skim milk Non-fat dry milk Dry or condensed buttermilk Frozen condensed milk Sweetened condensed milk Dry milk
Whey solids	Cheaper source of serum solids	Reduces shelf-life of ice cream due to high lactose content Lower protein content reduces protein content of ice cream Legal limit of 25% replacement of total serum solids in mix	Dry sweet whey Dry acid whey (in limited quantity) Condensed whey

improves the keeping quality over that of plain condensed milk. With this concentration of sugar, the osmotic pressure of the solution is high enough to suppress the growth of practically all microorganisms. The product will keep at room temperature. However, sweetened condensed products have a cost disadvantage and are not very popular in the ice cream industry. The freight cost of shipping sugar in the form of sweetened condensed milk is more than the shipping cost of dry bagged sugar.

Other condensed milks. Using frozen condensed whole or skim milk is a fairly expensive way to store solids. These ingredients are therefore not in general use. Evaporated milk is also not generally used because of the mandatory sterilization treatment required in its production. This heat treatment imparts a noticeable cooked flavor and a carmelized color.

The titratable acidity test (see Chapter 3, Box 2) should be applied to all condensed milk products. When diluted so as to contain the same MSNF concentration as skim milk, the acidity should be approximately that of fresh skim milk (0.18%).

Dry buttermilk and condensed buttermilk. Sweet-cream buttermilk is obtained from churning cream that has not developed noticeable acidity. Buttermilk has beneficial effects on the whipping ability of the mix because of its inherent lecithin content. It also contributes to richness of flavor. Buttermilk lipids tend to deteriorate readily in storage, so care must be taken to use only fresh ingredients. The fat content of condensed dry buttermilk is 3–4% and should be taken into account when formulating the ice cream mix.

Nonfat dry milk solids. NFDM is one of the most concentrated sources of serum solids and is frequently used. Spray-dried powder should have good flavor and light color and should be free from scorched particles and easily dispersible. NFDM should be bought only in such quantity as can be used before the product develops off-flavors. It should preferably be kept in cold storage. This slows down the development of stale flavor, which can impart an old or "storage" flavor defect to ice cream. NFDM offers an advantage over fluid products because it is the most concentrated low-cost source of serum solids and can be stored up to one year without loss in quality. For use in an ice cream mix, extra-grade low- or medium-heat NFDM is recommended. There is no great advantage in using more expensive agglomerated or instant NFDM for frozen desserts. Incorporation of dry milk into an ice cream mix is facilitated by the use of a powder funnel and blender pump or special blending equipment.

Whey. Dry sweet whey is commonly used to replace 25% of the serum solids in ice cream. Because dry whey contains 72% lactose as compared to 52% lactose in serum solids, its incorporation beyond this level may crystallize lactose to cause a "sandiness" defect in the ice cream. Delactosed whey, demineralized whey, and whey protein concentrates are not frequently used as they are relatively expensive and offer few functional advantages.

Sodium and calcium caseinates. These improve whipping properties of the mix and heat-shock resistance, body, and texture of the ice cream. Sodium caseinate at about 0.5% in the mix accords a slow-melt character to ice cream. However, these ingedients are relatively expensive and very rarely used.

Hydrolyzed proteins. Hydrolyzed milk proteins may be derived from the casein or whey fraction of milk and act as functional stabilizers in frozen desserts. They can replace gelatin, gums, celluloses, alginates, and other hydrocolloids. Although expensive, these ingredients may be preferred for their consumer-friendly appearance on the label. Hydrolyzed whey proteins are subject to a limitation of 25% serum solids replacement.

Reduced-lactose products. To improve heat-shock resistance and avoid the "sandiness" defect of frozen desserts, reduced-lactose concentrated milk can be used as a source of serum solids. The lactose-reduced ingredient is manufactured by holding the product at room temperature to bring about lactose crystallization (after a condensing step in a vacuum pan). Crystallized lactose is centrifuged out to obtain the reduced-lactose product. It has been used in high-solids ice cream mixes. Approximately 25% of serum solids can be replaced with low-lactose skim milk.

Another approach for lactose reduction is to use food-grade lactase to effect hydrolysis of lactose before the condensing step. In some cases, ultrafiltration of skim milk to obtain a retentate of reduced lactose content has been employed. Removal of lactose from skim milk increases the protein content, which, in turn, increases the acidity and viscosity of the mix. However, the texture and storage quality of the frozen dessert are markedly improved. Ice cream mix containing lactose-hydrolyzed serum solids freezes at a lower temperature and does not become too hard in storage. Because of the conversion of the disaccharide lactose to the monosaccharides glucose and galactose, a noticeable depression in freezing point is brought about. Accordingly, at *draw* and storage temperature, the ice cream displays a noticeably softer texture, which is a feature of interest to consumers. Lactose hydrolysis, therefore, provides a tool to favorably alter the melting properties of ice cream. Use of reduced-lactose serum solids, on the other hand, leads to no difference in melting characteristics of the frozen dessert.

Lactose hydrolysis also affects the sweetener level. Lactose is only 16% as sweet as sucrose, but lactose-*hydrolyzed* products are much sweeter (53% as sweet as sucrose). Thus, the sweetener content of the mix can be reduced a

Draw temperature—The temperature at which frozen dairy products are removed from the freezer for optimum body and texture and to facilitate packaging.

Hydrolysis—Enzymatic cleavage of glyceride molecules or proteins to liberate their constituent fatty acids or amino acids, respectively.

TABLE 4-8. Typical Formulation of Various Grades of Hard-Frozen Ice Cream

Component	Economy	Regular or Brand Name	Deluxe or Super-Premium
Milkfat, %	10	10–12	16–18
Nonfat milk solids, %	7.5	7.5–10	7–9
Whey solids, %	2.5	0–2.5	0
Sucrose, %	0	6–12	15.5
55 DE[a] high-fructose corn syrup, %	7.6–9.0	0–6	0
36 DE corn syrup solids, %	9–11.4	6	0
Stabilizer/emulsifier, %	0.40	0.35	0–0.35
Total solids, %	37–40	38–40	40–42
Weight/gal, lb	4.5	4.5	5.5–6

[a] Dextrose equivalent.

TABLE 4-9. Comparative Formulation of Soft-Serve and Hard-Pack Forms of Low-Fat and Nonfat Ice Cream Mixes

Component	Soft Serve		Hard Frozen	
	Nonfat	Low-Fat	Nonfat	Low-Fat
Milkfat, %	0–0.5	4–7	0–0.5	4–7
Nonfat milk solids, %	12–14	13.5–15	12–14	13–15
Whey solids, %	3–3.5	4.5–5	3–3.5	4.5–5
Sweeteners (sucrose, corn sugar), %	13	13	15	15
Stabilizers/emulsifier, %	An[a]	An[a]	An[a]	An[a]
Total solids, %	30–32	32–34	35–36	36–38

[a] As needed.

little when lactose-hydrolyzed skim milk is used.

From the standpoint of formulation, various grades (or lines) of ice cream are obtained by changes in the type and concentration of the various ingredients described above (Table 4-8).

SOFT FROZEN DAIRY PRODUCTS

These products constitute soft-serve ice cream, which is served immediately from the ice cream freezer. Milk shakes and smoothies also belong to this category. These products are generally lower in fat than their hard-pack counterparts. Fat substitutes based on starch, pectins, and microparticulated whey proteins along with gums, cellulose gel, microcrystalline cellulose, maltodextrins, sodium caseinate, etc., provide body and texture to the product. The serum solids content varies from 10 to 16%, and the total solids vary from 30 to 35%. In comparison with hard ice cream, soft frozen desserts contain higher serum solids and lower sweetener levels. The draw temperature is also lower (–7.8 to –6.7°C, 18–20°F) for soft-serve products. Formulation of reduced-fat frozen desserts is shown in Table 4-9.

Milk shakes can be formulated to give certain consumer-desired attributes. They may be cold and refreshing, smooth and slick, thick and viscous, coarse and icy, or light. The composition of two main types of shakes is shown in Table 4-10.

References

1. Kosikowski, F. V., and Mistry, V. V. 1997. *Cheese and Fermented Milk Foods*, 3rd ed. Vols. 1 and 2. L. L. C. Kosikowski, Westport, CT.

2. U. S. Food and Drug Administration. *Code of Federal Regulations*. Part 135. Frozen desserts. April 15, 1995.

3. Marshall, R. T., and Arbuckle, W. S. 1996. *Ice Cream*, 5th ed. Chapman & Hall, New York.

4. Jimenez-Flores, R., Klipfel, N. J., and Tobias, J. 1992. Ice cream and frozen desserts. In: *Dairy Science and Technology Handbook*, Vol. 2. *Product Manufacturing*. Y. H. Hui, Ed. VCH Publishers, Inc., New York.

TABLE 4-10. Typical Formulas for Milk Shakes

Composition	Direct-Draw Shake[a]		Spindle-Flavored Shake[b]	
	Smooth and Thick	Coarse and Icy	Smooth and Thick	Coarse and Icy
Milkfat, %	0–4	0–4	0–4	0–4
Milk solids-not-fat, %	12	10	12.5	11.5
Sweetener (sucrose equivalent), %	12	12.5	9	8
Stabilizer/emulsifier, %	0.5	0.2	0.35	0.2

[a] Draw temperature: –5 to –4.4°C (23–24°F), 50% overrun.
[b] 1–2 oz of syrup added at spindle stage contributes 5–7% sweetener (sucrose equivalent) in a 16-oz shake. Draw temperature: –5.6 to –5°C (22–23°F).

Baked Products

Ingredients used in baking consist primarily of flours, sugars and syrups, shortening, milk products and blends, eggs and egg products, and water. Flours, sugars, water, and shortenings are major and essential components of most bakery items. Dairy-derived ingredients (including butter), eggs, and other minor ingredients contribute desirable properties to the products.

Ideally, whole milk or its concentrates would be used to produce baked goods with premium flavor. In reality, coproducts from the dairy industry are commonly used, primarily to satisfy the economic conditions of commercial bakery businesses. Bakery blends based on nonfat dry milk (NFDM), whey solids, whey protein concentrates, and sodium or calcium caseinates constitute valuable ingredients in bakery products and are available for customized use by the baker. These ingredients, which are used along with defatted soya flour, emulsifiers, and dough conditioning agents, provide nutritional value as well as desirable functional properties. Economical substitute blends with functional attributes comparable to those of NFDM have also been developed.

Functional milk constituents (other than milkfat) of interest to bakers are milk sugar and milk protein (1). These contribute to the structure of the final product, to flavor and color development through browning reactions, and to texture, grain, and crumb structure. They tenderize the product, promote moisture retention, extend shelf life, and supply a food source for yeast during fermentation. Dry and liquid whey are economical sources of lactose, whereas NFDM, dry buttermilk, and whey protein concentrate provide the functionality of lactose and milk protein. The baking quality of NFDM can be assessed by a standard procedure (2).

Whey and whey products have been used in many bakery applications. They promote color formation in crust and crumb, increase volume, and improve texture. Caramelization of lactose during baking gives characteristic aromas and flavors. Specialty whey products can replace egg products at considerable savings in ingredient cost. Examples of dairy products in some bread types are given in Table 5-1.

Butter in Baked Products

The selection of a particular fat for use in formulating baked goods is based on its performance during manufacture of the bakery items and the attributes imparted. Real and perceived health concerns also dictate the selection of fats. The desire to reduce saturated fat and increase unsaturated fat in the diet has prompted the use of certain fats with lower levels of saturated fatty acids (3). The composition of some bakery fats is shown in Table 5-2. The contributions of fats to baked products are listed in Box 5-1.

To fulfill a need for enhanced functionality in milkfat, the dairy industry has developed fractionated and plasticized butterfat specifically for pastry, cookie, and cake manufacture. A low-melting glyceride (LMG) fraction prevents the appearance of fat crystals (fat bloom defect) during the shelf life of short bread cookies made with butter. The LMG fraction also traps a higher concentration of lactone and methyl lactone flavor precursors. Therefore, it assists in developing more intense butter flavor in the product. Butter made from anhydrous milkfat enriched with high-melting glycerides duplicates the functionality of nondairy pastry fat in croissants and Danish pastry. The ingredient cost is considerably moderated by blending vegetable oil with butterfat in a near 1:1 ratio. Butter in powdered form is also available for use in dry bakery mixes.

FUNCTIONS OF BUTTER

Butter performs a variety of functions in baked products. It contributes to flavor, mouthfeel, texture, shelf life, and tenderness.

Flavor. In addition to imparting "upscale/gourmet" image, butter gives a distinctive flavor to biscuits, breads, frostings, icings, buttercreams, cookies, crackers, croissants, muffins, cakes, pastries, and pie crusts. It is also a good carrier of flavor throughout the product, assisting in the uniform distribution of flavorings. Reaction flavor compounds (e.g., lactones) are generated from the fatty acids of butterfat during baking. Enzyme-modified butterfat products can be

TABLE 5-1. Composition of Dough for Some Types of Bread

Ingredient	Baker's Percent[a,b]			
	Pan Bread	Country Style	Milk Bread	Milk Rolls
Flour	100	100	100	100
Sweetener solids	7.25	5.0	7.0	10.0
Yeast	2.75	2.5	3.0	5
Yeast food	0.5	...	0.375	0.5
Salt	2.1	2.5	2	2
Shortening	2.3	2	...	4
Honey	...	5	0	0
Butter	...	3.5	3.5	3.5
Dairy blends/ nonfat dry milk	2	8	6	6
Emulsifier	0.50	...	...	...
Dough conditioner	0.50	...	...	...
Protease enzyme	0.25	...	...	...
Calcium acid phosphate	...	...	0.25	0.25
Preservative	0.20	...	...	...
Water	64	66	66	66
Malt	...	...	0.5	0.5

[a] Adapted from (8,9).
[b] Percent flour basis.

TABLE 5-2. Composition of Some Bakery Fats[a]

Source of Fat	Fat Content (%, wt/wt)	Saturated Fat (%)	Monounsaturated Fat (%)	Polyunsaturated Fat (%)
Butter	80	50.2	23.4	3.0
Soft margarine	80	13.8	28.5	34.6
Vegetable shortening	100	25.0	44.5	26.1
Lard	100	39.2	45.1	11.2

[a] From (4); used by permission.

used to give intense flavor notes of butter, allowing butter to be used at low levels to meet the consumer demand for reduced-calorie and reduced-fat products.

Lecithin, which is naturally present in butter at a 0.25% level, is instrumental in stabilizing fat emulsions, leading to uniformity of texture, aroma, and flavor in baked foods. In this regard, butter and buttermilk perform better than anhydrous milkfat, which contains essentially no lecithin.

Mouthfeel. Butter imparts highly desirable mouthfeel characteristics to baked products. This property is related to the ratio of the crystalline (solid) and noncrystalline (liquid) fractions of butter as a function of temperature. This solid fat index (SFI) is related to the percent solid fat in a fat (Table 5-3) (4–7). More recently, a measurement of solid fat content based on pulsed nuclear magnetic resonance has been developed.

Butter exhibits a steep and sharp SFI curve along with a low melting-point range of 27.8–37.2°C (82–99°F). It melts completely at 37.8°C (100°F). At body temperature, only 5% is in solid form, and therefore butter gives no "waxy" mouthfeel. It is commonly combined with some margarine and shortening products to alleviate the waxy mouthfeel observed when only shortenings with a relatively broad SFI curve (e.g., puff pastry margarine) are used.

Flakiness and tenderness. Butter serves a useful function in imparting flakiness to croissants and Danish and puff pastry. For pies and tarts, flakiness is increased by mixing part of the butter into the dough. Flakiness is caused by entrapped carbon dioxide. In comparison with lard and some hydrogenated vegetable oils, butter does not possess good shortening properties. However, development of a fine crystal structure in butter helps to impart better shortening properties to cakes and cookies.

Shelf life. Like other shortenings, milkfat from butter, butteroil, etc. provides a barrier to loss of moisture after baking. Furthermore, it is claimed to inhibit retrogradation of starch (which is associated with staling). Thus, tenderness and flakiness are maintained during the shelf life of baked goods. This effect on shelf life is

Box 5-1. Contributions of Fats in Baked Products

- Tenderization of products by preventing formation of three-dimensional starch-protein network.
- Texture modification of dough and baked goods, e.g., shortening effect.
- Aeration during mixing.
- Extension of shelf life by slowing retrogradation of starches (antistaling effect).
- Modification of dough consistency.
- Heat transfer.
- Flavor carrier.
- Emulsification.
- Release from pans.
- Flakiness in laminated bakery items.
- Interaction with gluten in yeast-raised doughs.
- Interaction with starch and modification of its hydration potential.

TABLE 5-3. Solid Fat Indices of Bakery Fats[a]

Source of Fat	Solid Content (%) at				
	10°C 50°F	21.2°C 70°F	26.7°C 80°F	33.3°C 92°F	37.8°C 100°F
Butter	32	12	9	3	0
Lard	25	20	12	4	2
Table margarine	28	16	12	3	0
Roll-in margarine	25	21	20	18	15
Puff margarine	28	25	24	22	19
All purpose shortening	33	28	22	10	8
Cake and icing shortening	28	23	22	18	15

[a] From (4); used by permission.

Creaming—High-speed mixing of a plastic shortening containing sugar in order to incorporate air.

TABLE 5-4. Croissant Formula and Processing[a,b]

Ingredient	Amount, g
Bread flour	1,860
High-gluten flour	1,240
Sugar, granulated	179
Salt	60
Nonfat dry milk	239
Water	1,922
Yeast	124
Roll-in butter	1,391[c]

[a] Processing steps: Using a mixer with 20-qt bowl and dough hook, mix dry ingredients at speed 1 for 1 min. Add water and yeast; mix at speed 1 for 1 min and at speed 2 for 3 min. Temperature should be 15.6°C (60°F). Remove the dough and preshape into rectangular form. Cover the dough and allow it to rest for 20 min. Sheet dough to 1/2-in. thickness. Apply roll-in butter to the surface of two-thirds of the dough sheet. Fold the sheet three times and allow it to rest for 20 min. Fold four times, rest for 30 min, followed by folding three times and retarding overnight. Bake for 14 min at 176.7°C (350°F).
[b] From (5); used by permission.
[c] 24%, dough basis.

observed in cakes as well as in yeast- or chemically leavened breads.

HANDLING OF BUTTER IN BAKERIES

Storage of butter in the original carton protects it from acquiring foreign flavors and aroma. It should be stored away from foods with strong odors and kept at 0–3.3°C (32–38°F) and 80–85% relative humidity for a maximum of four months. Butter can be held for a year in frozen form (–23.3 to –28.9°C, –10 to –20°F).

Butter should be tempered to an appropriate temperature for optimum functionality. Softening at 15.6–18.3°C (60–65°F) is advisable. The recommended thawing time is two to three days for a 30-lb case and four to five days for a 68-lb block. Acceleration of thawing by subjecting butter to higher temperatures changes the crystal profile and should be avoided. Liquefied warm butter (about 37.8–43.3°C [100–110°F]) gives a greasy dough with undesirable viscosity or density and may compromise the quality of the baked product. For pie and pastry dough, cold butter works well. Before the addition of flour, sugar, and other ingredients, cold butter (to be used as a roll-in fat) can be made pliable by beating it with a hook or paddle in a mixer. In the production of cookies, cakes, breads, and icings, butter is warmed to room temperature. For cake batters, sugar is thoroughly *creamed* with butter at room temperature to achieve a uniform distribution in the batter or dough. Antioxidants (BHA and BHT) at a level of 0.005% may be added to baked foods with long shelf life (e.g., cookies) to avoid oxidative deterioration of the fat. Use of butter in the production of croissants is shown in Table 5-4.

Hard Wheat Products

PASTRY AND LAMINATED DOUGHS

For a laminated dough, the dough is rolled out and butterfat is applied. The dough is then folded and rerolled and more butterfat is applied. After several repetitions, several layers of butterfat are interspersed with layers of dough. When the pastry is baked, moisture in the butter turns into steam. Carbon dioxide is released from the leavening, making the dough layers rise. The rise (or volume) of baked pastry is directly proportional to the percentage of solid fat.

Research has shown that *European-style cultured butter* containing no salt or coloring and a minimum of 82% butterfat performs better than regular butter. Croissants made with European-style butter had higher specific volume and better flakiness inside and outside (4).

BREAD AND BISCUITS

Butter contributes tenderness to bread by interfering with the development of the gluten network. This may result in extended shelf life. Butter can be used as a carrier of herbal flavor and gives an attractive appearance to bread when sprayed on the loaf surface. Butter can also contribute premium flavor, clean mouthfeel, and flakiness to biscuits.

In addition to the contributions of butter, milk products also contribute to flavor and color, crust and grain development, and loaf volume. Nonfat milk solids influence water absorption in dough, mixing requirements, yeast fermentation rate, bromate response in flour, baking time and temperature, and physical properties of the bread. Up to 6% NFDM, based on flour weight, has historically been incorporated to simulate the use of skim milk for all liquid ingredients in the formula. NFDM tends to strengthen dough, thereby producing longer mixing time. Hydration of milk protein during fermentation "tightens" the dough, and milk proteins compete with flour for available water. To prevent lumps in the dough mix, dairy blends are generally scaled on top of flour in the dough mixer.

Fermentation time as well as fermentation tolerance are enhanced by the inclusion of NFDM in bread formulas. Milk products provide enzyme cofactors such as ammonium ions for yeast metabolism during fermentation. They also serve as food for yeast, and increased yeast activity can translate to increased loaf volume. Fermentation tolerance is affected favorably, resulting in bread of uniform quality (volume, flavor, crumb firmness, crust and crumb color, and shelf life) from batch to batch.

In the sponge-and-dough process, dry milk is generally added at the dough stage (8). It can also be incorporated into the sponge to compensate for a flour's low protein content, excessive amylolytic activity, short fermentation toler-

European cultured butter— Butter made from cream to which a lactic culture has been added. It contains no salt and has a characteristic diacetyl flavor.

TABLE 5-5. Formulation and Processing[a] of Full-Fat and Reduced-Fat White Cake[b]

Ingredient	Formula (Baker's Percent)	
	Full-Fat Cake	Low-Fat Cake
Cake flour	100.0	100.0
Granulated sugar	130.62	130.97
Water	106.21	112.63
Liquid egg yolks	48.27	0.0
Liquid egg whites	0.0	45.04
SB fat replacer[c]	0.0	15.34
Nonfat dry milk	8.95	8.96
Salt	2.57	2.53
Baking powder	6.12	6.17
Vanilla flavor	0.0	0.85
Lecithin	0.26	0.0
Emulsified shortening	25.27	0.0
Total	**428.27**	**422.49**
Specific gravity	0.70	0.73
Total cake volume index[d]	552	550
Cake score (maximum: 100)	94	92
Fat per 80-g serving, g	7.8	1.5

[a] Processing: Thoroughly mix dry ingredients together. Blend dry mix, water, and egg whites in mixer for 1 min on low speed. Mix for 1 min on low, then 3 min on medium speed. Bake at 176.7°C (350°F) for 20–27 min.

[b] From (3); used by permission.

[c] K-Blazer.

[d] Calculation based on top diameter plus bottom diameter. Bottom diameter includes measurements taken horizontally (bottom) and vertically (five places).

TABLE 5-6. Formulation and Processing Suggestions for Cracker Production[a,b]

Ingredient	Weight (g)	Baker's Percent
Flour, bread	317.50	100.00
Salt	6.33	1.99
Baking powder	2.06	0.65
Butter	73.36	23.10
Milk, whole	113.00	35.60
Egg	57.00	17.95
Total	569.25	179.29

[a] From (5); used by permission.

[b] Yield: approximately 60 crackers. Procedure: Sift the flour, salt, and baking powder into a bowl. Add the butter, milk, and egg; mix to make a stiff dough. Knead and roll the dough very thin (about 1/8-in. thick). Cut into squares or rounds and place on parchment-lined baking sheets. Prick the crackers with a fork. Bake in a 204.4°C (400°F) oven for 10 min or until lightly browned.

Amphoteric—Describing a compound such as a protein that has both positive and negative charges.

High-ratio—Describing a cake that contains more sugar than flour.

ance, and poor strength. Dough containing NFDM is set at a higher temperature, e.g., 27.2°C (81°F) in winter and 25.6–26.1°C (78–79°F) in summer. It requires 2–3 min longer for recovery in the overhead proofer as well as in the pan. The proof box temperature should not exceed 37.8°C (100°F), and the relative humidity should be slightly lower (9).

The no-time bread dough method does not require long bulk fermentation for conditioning the wheat gluten. Oxidizing agents (bromate, ascorbic acid, or azodicarbonamide) are added to condition or strengthen the dough. The addition of 4–6% NFDM to no-time dough was shown to produce bread quality equivalent to that of bread derived from the conventional sponge-and-dough process (10). The *amphoteric* nature of the milk proteins allows them to act as a buffer against pH changes, offering protection against overoxidation by oxidizing agents, which negatively affects loaf volume and crumb grain.

Soft Wheat Products

CAKES

Whole milk, if added to cake batter, primarily provides moisture (water). The solids contribute functional components: lactose and protein. As a reducing sugar, lactose contributes to the browning of the cake crust during the later stages of baking. This function is relatively unimportant in cakes to be iced or those with normally dark crusts (e.g., chocolate); however, it is important in cakes that are not fully frosted. Caramelization of lactose also contributes to the characteristic aroma and flavor. It has been claimed that, because lactose takes part in the browning reaction, the production of melanoidins is increased to the point where cake flavor is affected. However, the actual contribution to cake flavor by this means is minimal, given the degree to which most retail (and wholesale) cakes are baked.

In cakes for which incorporation of air nuclei in the aqueous phase of the batter is important (e.g., *high-ratio* and/or Oakes or continuous-mix types), the soluble protein in milk can aid in trapping air during mixing.

Whey protein products have good emulsifying quality in high-fat bakery items because they contain both hydrophobic and hydrophilic groups. Consequently, they help distribute fat in the dough, which can lead to a more finely divided (smooth or silky) crumb and possibly a slightly higher finished volume. However, because these proteins are not set by heat, they do not contribute to the finished cake structure. Whey proteins denature at baking temperatures and exhibit good water-binding ability. This contributes to moistness, tenderness, and enhanced keeping quality. Consequently, milk can be considered an ingredient that tenderizes the crumb of the cake, albeit in an indirect manner. There is some indication that not all NFDM preparations are equally able to function in this way.

Specialty whey products can replace egg products at considerable cost savings. Milk products can affect the solids-liquids balance (often referred to as formula balance), which is critical to achieving optimum batter performance. Formula balance is covered in basic terms in reference 8. A formula for full-fat and low-fat cake is shown in Table 5-5.

CRACKERS

Cracker doughs are generally stiff because of their low moisture content. Their stiffness interferes with the lamination and sheeting into thin layers that are necessary to produce cracker texture. Butterfat reduces this stiffness, making the dough more plastic for lamination. Sodium bisulfate hydrolyzes the disulfide bridges of gluten, thereby further facilitating lamination. Butteroil can be sprayed on cracker surfaces to enhance gloss and appearance. A formula and processing steps for crackers are shown in Table 5-6.

OTHER PRODUCTS

In the manufacture of muffins, NFDM gives structure and body to the product. It is also involved in flavor development, moisture retention, and extension of shelf life. Muffin formulas typically use 5–12% NFDM based on flour at 100%. A suggested formula for dry muffin mix contains 24.4% sugar, 15.0% shortening, 2.8% egg solids, 0.5% salt, 3.0% NFDM, 52.5% cake flour, 1.8% baking powder (11).

Dry sweet whey is widely used in bakery items, especially cookies, because of its superior functionality with respect to color and texture formation in the finished product. Butter, used in butter cookies, makes the cookies moist and soft as a result of the interaction of butterfat, other solids, air, and water. Butter also pro-

TABLE 5-7. Fat Reduction in Sweet Dough[a]

Ingredient	Full-Fat		Low-Fat[b]	
	Actual (%)	Baker's (%)	Actual (%)	Baker's (%)
Bread flour	37.27	78.8	31.32	58.7
Pastry flour	10.00	21.2	22.00	41.3
Sugar	8.20	17.4	5.50	10.3
Corn syrup solids	...	...	2.00	3.8
Vegetable shortening	8.20	17.4	...	...
Nonfat dry milk	2.50	5.3	...	...
Salt	1.00	2.1	0.95	1.8
Flavors	0.08	0.2	0.08	0.2
Emulsifiers	0.95	2.1	0.95	1.8
Whole eggs	9.00	19.0	...	...
Color (FD&C yellow #5 & #6)	...	...	0.002	0.004
Oil	...	...	1.00	1.90
Yeast (instant)	1.30	2.8	1.30	2.4
Water	21.50	45.5	27.40	51.4
Simplesse 100 (liquid)	...	...	7.50	14.1
Total	**100.00**	**211.8**	**100.00**	**187.7**
Fat, %	11.4	3.0		
Fat reduction, %	...	73		
Calories from fat, %	34	9		

[a] From (12); used by permission.

[b] Processing of low-fat sweet dough: Combine all dry ingredients except color (including yeast), using a paddle on low speed. Mix until well blended. Add oil gradually while mixer is on low speed. Mix until uniform. Combine water, Simplesse, and color. Add to dry ingredients. Mix on low speed, using a dough hook, until dough absorbs all the water, approximately 2 min. Mix on medium speed 4–6 min, until full gluten development is reached. Scrape down bowl and dough hook halfway through mix time. Shape dough into a ball and flour the dough surface. Let rest 5–10 min. Portion into desired shape. Proof at 80–90% relative humidity and 43.3–46.1°C (110–115°F), approximately 30 min. Bake in 162.8°C (325°F) preheated oven 10–25 min or until golden brown. Bake times will vary depending on dough weight.

TABLE 5-8. Formulation and Processing[a] of Full-Fat and Low-Fat Cinnamon Rolls[b]

Ingredient	Formula (Baker's Percent)	
	Full-Fat Rolls	Low-Fat Rolls
Sponge		
Bread flour	74.7	74.7
Water	58.2	58.2
Yeast (compressed)	3.02	3.0
Powdered egg whites	0.0	2.0
Vital wheat gluten	2.0	2.0
Yeast food	0.8	0.8
Powdered whole eggs	3.1	0.0
Dough		
Raisins	30.4	30.4
Bread flour	25.3	25.3
Granulated sugar	9.1	9.1
Dextrose (glucose)	9.1	9.1
Water	8.1	8.1
High-fructose corn syrup	7.1	7.1
SB fat replacer[c]	0.0	5.1
Non-fat dry milk	4.0	4.0
Salt	1.8	1.8
Yeast (compressed)	1.8	1.8
Sodium stearoyl lactylate	0.3	0.3
Mono- and diglycerides	0.3	0.0
Vegetable shortening	15.2	0.0
Filling		
Sugar	19.0	19.0
Cinnamon	1.9	1.9
Total	**275.0**	**263.9**
Fat per 55-g serving, g	3.6	0.8

[a] Processing: Mix all sponge ingredients 1 min on low and then 3 min on medium speed. Ferment (using proof box) for 3 hr. Dry blend all ingredients for dough part. Add sponge and water; mix with dough hook 4 min on low, then 6.5 min on medium, followed by 1.5 min on high speed. Allow dough to rest for 30 min. Roll out dough to approx. 3/8 in. thickness. Sprinkle on cinnamon and sugar. Roll up and cut into desired weights. Proof for 30–35 min. Bake for 15 min at 190.6°C (375°F).
[b] From (3); used by permission.
[c] K-Blazer.

DATEM—Diacetyl tartaric acid esters of monglycerides, an emulsifier.

vides desirable flavor and mouthfeel to cookies. Enzyme-modified butter flavor can be used in cookies to furnish butter flavor with reduced calories.

The use of butter and anhydrous milkfat gives a light texture to icings and creams. Liquid fat, along with the solid fat fraction, creates a kinetic barrier by enveloping air cells, thereby stabilizing foam or air cell structure. Milkfat tempers sweetness in butter creams and icings.

Role of Dairy Ingredients in Fat Reduction of Baked Foods

Currently, fat reduction in bakery items is a priority, instituted in response to consumer demand. Fat-replacing ingredients are based on whey and milk protein concentrates, starch and starch derivatives, hydrocolloids, emulsifiers, celluloses, and synthetic fats (7,12,13). Simplesse, a commercial microparticulated whey protein concentrate, can be used to replace fat because of its fatlike sensory characteristics in a range of bakery items (12). The functional properties of fat in baked goods relate to appearance, flavor, texture, and mouthfeel. No single ingredient delivers all the functionality of fat. Therefore, several ingredients are used to develop reduced fat-products.

Table 5-7 shows a formula and procedures for reduced-fat sweet dough using Simplesse.

Another commercially available fat replacer for baked goods is called SB fat (trade name, K-Blazer) which consists of whey, propylene glycol monoester, modified food starch, oat fiber, mono-and diglycerides, polysorbate 60, sodium caseinate, *DATEM* emulsifier, potassium phosphate, and xanthan gum. It replaces shortening and whole eggs in formulations. Its use in low-fat cinnamon rolls is illustrated in Table 5-8.

Troubleshooting

PASTRY/LAMINATED DOUGHS		
Symptom	**Causes**	**Changes to Make**
Low volume	Insufficient number of layers in final dough	Increase number of sheeting steps.
	Insufficient roll-in fat between layers	Apply more roll-in fat during processing. Check for oiling out.
	Dough too stiff	Increase retarder temperature or increase proofing time.
Oiling out	Too much roll-in fat	Reduce amount of roll-in added.
	Dough and/or roll-in fat temperatures too high	Lower retarder temperature. Return dough-fat to retarder between sheeting steps.
	Dough too stiff	Increase retarder temperature or increase proofing time.
	Excessive roll pressure	Reduce pressure and increase number of passes through sheeters.
Coarse texture	Thick layers	Reduce sheeting rate and increase number of passes. Increase proofing or retarding time.
	Fat breaks through dough	Adjust amount or temperature of roll-in fat.
Dough tearing during roll-in	Roll-in fat much harder than dough	Temper roll-in fat at application temperature before use. Use roll-in fat with a lower SFI profile.
	Dough too soft	Decrease dough absorption. Make sure dough is adequately retarded before rolling in.
Loss of flakiness	Roll-in fat completely melting in proofbox	Lower proofbox temperature.
	Roll-in fat too soft	Return dough-fat book to retarder between sheeting steps. Use roll-in fat with a lower SFI profile.

BREADS/BISCUITS/ROLLS

Symptom	Causes	Changes to Make
Dark crust color	Excessive browning	Reduce lactose levels. Increase baking time and decrease baking temperature.
Poor texture	Absorption too low or protein hydration insufficient	Add more water. Increase mixing time. Add monocalcium phosphate.
Clumping of milk powder during mixing	Inadequate hydration	Use instantized or agglomerated powders. Preblend flour and milk powder or use more rigorous mixing step.
Caking of powder	Improper storage conditions	Check packaging and store under cool and dry conditions.
	Age of powder	Check storage conditions and length of storage period.

CAKES

Symptom	Causes	Changes to Make
Dark crust color	Excessive browning reaction	Reduce lactose levels. Increase baking time and decrease baking temperature.
Low volume	Insufficient protein hydration	Increase water content or decrease level of milk proteins. Use powders exposed to high-heat treatment.
Clumping of milk powder	Inadequate hydration	Use instantized or agglomerated powders. Preblend flour and milk powder or use more vigorous mixing method.
Low volume with dip in center	Inadequate emulsification	Increase mixing or increase emulsifying agents.
	Poor foam stability (in oil cakes)	Add 1–3% soluble protein source.

References

1. Cocup, R. O., and Sanderson, W. B. 1987. Functionality of dairy ingredients in bakery products. Food Technol. 41(10):86-90.
2. American Association of Cereal Chemists. 1995. *Approved Methods of the AACC*, 9th ed. Method 10-85. The Association, St. Paul, MN.
3. Abboud, A. 1995. Technical Bulletin: Systems Approach to Reducing Fat in Baked Goods, 17(12). American Institute of Baking, Manhattan, KS.
4. Smith, R. 1993. Technical Bulletin: Use of European Style Butter in Bakery Foods, 15(10). American Institute of Baking, Manhattan, KS.
5. Anonymous. 1995. Dairy Ingredient Application Guide. Butter and Other Milkfat Ingredients. Dairy Management Inc., Rosemont, IL.
6. Matz, S. 1996. *Ingredients for Bakers*, 2nd ed. Pan Tech International, Inc., McAllen, TX.
7. Stauffer, C. E. 1996. *Fats and Oils*. American Association of Cereal Chemists, St. Paul, MN.
8. Pyler, E. J. 1992. Baking science and technology. In: *Encyclopedia of Food Science and Technology*, Vol. 1. Y. H. Hui, Ed., John Wiley, New York.
9. American Dry Milk Institute. 1968. *Nonfat Dry Milk in Cakes and Other Bakery Foods*. The Institute, Chicago.
10. Doerry, W. 1989. Technical Bulletin: Nonfat Dry Milk in No-Time Breads Doughs, 11(4). American Institute of Baking, Manhattan, KS.
11. Schierioth, W. 1991. Technical Bulletin: Scaling Up for Baked Product Development, 13(10). American Institute of Baking, Manhattan, KS.
12. Corliss, G. A. 1992. Technical Bulletin: Protein-Based Fat Substitutes in Bakery Foods, 14(10). American Institute of Baking, Manhattan, KS.
13. Sherwin, C. 1995. Technical Bulletin: Use of Whey and Whey Products in Baked Goods, 17(11). American Institute of Baking, Manhattan, KS.

Chocolate and Confectionery Products

Confections belong to two categories depending on their crystalline or amorphous structure (1–3). Crystalline confectionery products include chocolate, creams, fudge, fondants, nougats, marshmallows, pralines, and pressed sweets or lozenges. The amorphous or noncrystalline products are hard or chewy caramels, taffy, toffee, and brittles. From the standpoint of dairy ingredient usage, milk chocolate, fudge, caramel, toffee, and pralines are the most significant.

Dairy-based ingredients used in confectionery products include whole milk powder, sweetened condensed milk, evaporated milk, *milk crumb*, butter, anhydrous butterfat, lactose, caseins, and whey products.

Large manufacturers tend to purchase fluid milk for conversion into condensed or dry milks by proprietary processes in their own plants. In general, this allows them economies of scale as well as control of quality and functional properties in their products for a competitive edge. Since confections are extremely low-moisture foods, small manufacturers prefer to purchase concentrated sources of milk solids.

Dairy-based ingredients are natural, traditional ingredients that contribute desirable flavor, texture, and color to the products. They enhance the nutritional value, particularly the protein and calcium profiles. Milkfat, milk proteins, and lactose impart desirable functionality either in their native form or by interaction with other confectionery ingredients (2,3). These ingredients have been described earlier. Their functional attributes are described in Table 6-1.

The functional properties of dairy ingredients must be compatible with the composition and process used in the confection manufacture. Confections may contain reducing sugars and high levels of sugars, starches, and fats. Processing conditions involve exposure to temperatures exceeding 100°C (212°F) for extended periods of time. Therefore, interactions among carbohydrates, proteins, and lipids of ingredients leading to formation of color and generation of flavor and texture in the finished product need to be balanced.

In This Chapter:

Chocolate Products
  Dairy-Based Ingredients
  Manufacturing
    Considerations

Confections
  Dairy-Based Ingredients
  Manufacturing
    Considerations

New Opportunities in Chocolate and Confections for Dairy-Based Ingredients

Troubleshooting

Milk crumb—An intermediate ingredient of milk chocolate, consisting of a mixture of milk solids, sugar, and chocolate liquor.

TABLE 6-1. Constituents of Dairy Ingredients and Their Functionality Attributes in Confections[a]

Constituent	Functional Contribution to Confections
Milkfat	Gives rich flavor, texture, and mouthfeel
	Provides lecithin and monoglycerides, which aid in emulsification
	Precursor of key flavor compounds formed by cooking, e.g., volatile fatty acids, lactones, ketones, aldehydes
	Acts as a flavor carrier
	Provides a moisture barrier
	Inhibits bloom defect
	Oxidation and/or hydrolysis produces distinctive desirable flavor compounds. (However, under certain conditions of processing and storage, off-flavors [rancid, fishy, sunlight, stale] may develop.)
Milk proteins	Aid in emulsification and miscibility of various ingredients
	Assist in whipping and foam formation in certain products
	Absorb moisture due to water-binding capacity
	With heat, can yield protein fiber network that gives rigidity, shape, texture, and chewability. Furthermore, impede movement of water molecules, leading to retardation of sugar crystallization.
	Contribute to favor and color development
	Improve nutritional profile
Lactose	Imparts chewiness and graininess
	Causes sandiness defects under certain conditions, e.g., at high lactose concentrations
	Acts as a flavor carrier
	Polymerizes to form a flexible and extensible matrix (textural effect)
	Caramelizes to generate color and flavor compounds (aldehydes, ketones, alcohols, acids, reductones, etc.) in caramels and toffees
	Provides reducing sugar moiety that can participate in Maillard browning, which affects color and flavor
	Acts as a precursor of pyrazines (nutty flavor), maltol and isomaltol (caramel flavor), furfurol and furfuraldehyde (cooked cereal odor)
	Contributes to dark color and bitter taste defects under certain conditions of overcooking

[a] Adapted from (2–5).

Chocolate Products

Chocolate confections constitute the largest segment of the confectionery industry. Chocolate is consumed as a snack in chocolate bars, as a topping, or as a coating on candy bars, ice cream, cookies, nuts, and fruits. The flavor of chocolate is basically derived from combining *chocolate liquor* with sugar and is accentuated by the addition of milk solids, lecithin, nuts, and other flavorings (6).

In the United States, milk chocolate accounts for nearly 60% of all chocolate products. Federal standards of identity regulate the permitted ingredients and their ranges. Generally, milk chocolate has a light color and contains 40–55% sugar. Sweet dark chocolate may

Chocolate liquor—The solid mass obtained when ground, liquefied cacao bean kernels are cooled.

contain 60–62% sugar, while bittersweet dark chocolate contains only 25–30% sugar. Confectioners coatings or *compound coatings*, which are less expensive analogs of milk chocolate, are subject to their own standards and specifications.

In compound coatings, a part of the cocoa butter is replaced with an appropriate vegetable fat. The coating is formulated specifically for use in enrobing ice cream bars, candy bars, and bakery items. The products are labeled "chocolate flavored" or "chocolatey" (6). The higher melting point of the vegetable fat is useful in designing coatings for use in semitropical or tropical climates. The sweetness sensation is tempered by the use of special flavorings to avoid a sickeningly sweet impact.

For the diabetic market, sugars are replaced with sorbitol, mannitol, xylitol, and/or high-intensity sweeteners like saccharin, aspartame, and acesulfame K. In fat-reduced categories, the fats are replaced with fat substitutes.

White (or pastel colored) chocolate contains no cocoa powder or liquor. It is made by blending cocoa butter, sugar, milk solids, and vanilla/vanillin flavorings. Table 6-2 illustrates formulations of chocolate products.

DAIRY-BASED INGREDIENTS

Dry whole milk, condensed milk, and milk crumb are the primary ingredients of choice. Malted milk and concentrated or dry buttermilk, which contribute dairy solids, are optional. Buttermilk imparts a characteristic flavor and provides additional levels of emulsifier (phospholipids). The usage level and types of buttermilk are defined in the *Code of Federal Regulations* of the U.S. Food and Drug Administration (FDA).

Anhydrous milkfat (or butteroil) and nonfat dry milk (NFDM) can be used, but the quality of the milk chocolate is not equivalent to that of milk chocolate derived from milk crumb or dry whole milk. Where regulations permit, demineralized whey and some whey products may be used as economical replacers of sucrose and chocolate liquor in standard chocolate products.

The functionality of milk solids for use in chocolate can be manipulated by processing treatments. As compared to spray-dried whole milk powder, the roller-dried ingredient is more functional in chocolate. Roller drying disrupts the milkfat globules, so most (90%) of the milkfat in roller-dried powder has been freed from the globule, whereas in spray-dried powder only 3–6% of the milkfat is not in globules. In case only spray-dried whole milk is available, it becomes necessary to incorporate an addi-

Compound coatings—Coatings containing fats other than cocoa butter but similar to regular chocolate in melting properties.

TABLE 6-2. Typical Formulations (%) for Chocolate Products[a]

Ingredient	Milk Chocolate	Dark Chocolate	Compound Coatings
Chocolate liquor	12.0	16.0	⋯
Cocoa butter	22.6	25.1	⋯
Cocoa powder, (10–12%) fat	⋯	⋯	8–12
Dry whole milk	20.0	⋯	⋯
Nonfat dry milk	⋯	⋯	8–10
Vegetable fat[b]	⋯	⋯	28–32
Sugar	45.0	58.4	45–56
Lecithin	0.3	0.5	0.3–0.5
Vanillin	0.1	0.03	0.03

[a] Adapted from (2,7).
[b] Melting point, 35–39°C (95–102°F).

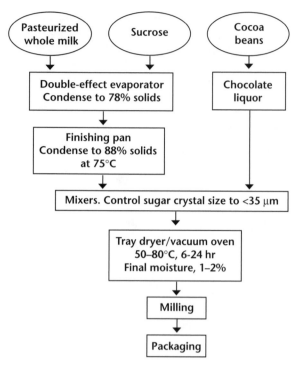

Fig. 6-1. Principle of milk crumb manufacture.

tional 2–3% cocoa butter to achieve equivalent handling properties during manufacture and to impart desirable mouthfeel (7,8). Milkfat is quite miscible with cocoa butter. Because cocoa butter is generally more expensive than milkfat, additional cocoa butter increases the cost.

Milk crumb is an intermediate ingredient in the manufacture of milk chocolate and chocolate coating. It has better keeping quality than whole milk powder. The typical composition of milk crumb is 16.5% total fat (9.2% milkfat, 7.3% cocoa fat), 6.2% fat-free cocoa, 53.5% sugar, 22.8% milk solids-not-fat, and 1.0% moisture. It is made from a blend of 13.5% cocoa liquor, 53.5% sugar, and 32% milk solids (1,2,8). Variations in milk solids are observed. Low-milk-solids crumb contains 13.5% cocoa liquor, 63% sugar, 4.5% cocoa butter, and 18% milk solids. High-milk-solids crumb contains 18% cocoa liquor, 55% sugar, and 26% milk solids (1). White crumb, which is used for white chocolate formulation, contains only milk solids and sugar.

The moisture content of 1–2% in milk crumb is a key factor in its superior shelf life. If milk crumb is exposed to atmospheric conditions, it picks up moisture from the environment and quickly develops off-flavors at a moisture level greater than 4%. Because of the heat treatment used in milk crumb manufacture, the bacteriological quality is excellent (yeast and mold count less than 50 CFU/g, total count less than 20,000 CFU/g). *Salmonella*, *Staphylococcus aureus*, and *Escherichia coli* are essentially absent.

MANUFACTURING CONSIDERATIONS

Milk crumb. The manufacture of milk crumb is shown in Figure 6-1. Some manufacturers mix water and whole milk powder to obtain reconstituted condensed milk to be converted into milk crumb. Compared to whole milk powder, milk crumb contributes a more pleasant flavor and smoother texture and mouthfeel characteristics in the final milk chocolate (4).

In the United States, chocolate manufacture entails the use of lipolyzed milkfat as such or in dry form in the formulation of milk crumb. Consequently, the milk crumb develops limited hydrolyzed milkfat flavor notes as well as toasted or caramelized flavor attributes. The short-chain fatty acids derived from hydrolysis of milkfat give rich flavor. The toasted flavor is ascribed to the generated flavor compounds furfurol and maltol.

Anhydrous butteroil is occasionally used at a 1–5% level to soften and standardize textural characteristics during different seasons. The FDA's standard of identity for milk chocolate does not permit caseinates, whey, or whey protein concentrates in milk chocolate. Lactose at the 0.001% level is used to control crystallization in milk crumb production.

Chocolate. Figure 6-2 illustrates the fundamentals of chocolate manufacturing.

The process involves mixing sugar, milk solids, chocolate liquor (also called baker's chocolate or unsweetened chocolate), and a portion of cocoa butter to form a semidry paste. The chocolate liquor is first melted by heating (with continuous stirring) in a jacketed vessel to 40°C (104°F). The hot water in the jacket should not exceed 60°C (140°F) to minimize burnt flavor defects. The ingredient paste is ground on a five-roll refiner to achieve particle size reduction. The ground mix is then transferred to a slow mixer (conche) to reduce particle size further. In this step, called conching, the fat layer softens. Conching takes several hours of mixing and blending, leading to increased viscosity caused by a decrease in particle size of the solid portion of the mix. After conching, the remaining cocoa butter and the lecithin and vanillin are incorporated. The melted chocolate is then poured into molds and allowed to solidify, followed by packaging (7).

Confections

DAIRY-BASED INGREDIENTS

Sweetened condensed milk, whole milk powder, and evaporated milk are used extensively in confection manufacture. Some manufacturers use reconstituted milk powder in place of con-

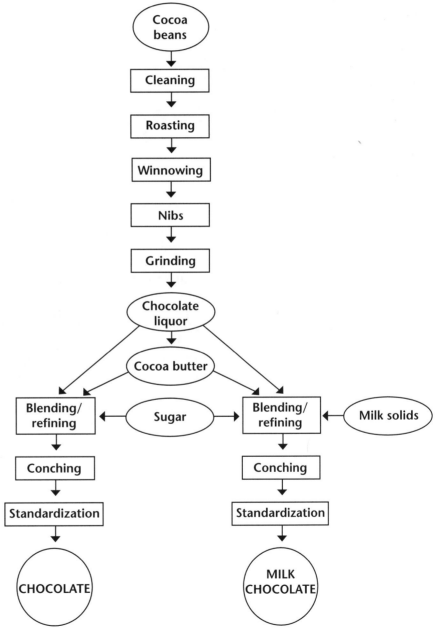

Fig. 6-2. Flow diagram for manufacture of chocolate and milk chocolate.

TABLE 6-3. Suggested Formulas (%) for Caramel, Fudge, and Toffee[a]

Ingredient	Caramel	Fudge	Toffee
Fondant	9.1	...	...
Sucrose	12.8	16.0	22.4
Brown sugar	12.8	...	...
Corn syrup solids[b]	21.9	40.5	33.7
Full-fat sweetened condensed milk	23.4	...	...
Nonfat dry milk	...	8.5	2.5
Vegetable fat[c]	10.3	11.5	11.3
Salt	0.4	0.5	0.4
Lecithin	...	...	0.3
Glycerin monostearate	0.6	...	...
Vanillin	0.1	0.1	0.1
Water	8.6	22.9	29.3
Total	100.0	100.0	100.0

[a] Adapted from (1).
[b] 42 dextrose equivalent.
[c] Melting point 32°C (90°F).

densed milk. To prepare the reconstituted milk, a mix consisting of 21.9% NFDM, 44.8% sucrose, 8% vegetable fat (melting point, 32°C), 0.4% lecithin, and 0.1% $NaHCO_3$ (pH 6.0–6.5) is prepared. NFDM and $NaHCO_3$ are mixed in cold water. After sugar addition, the mix is heated to 70°C (158°F). The lecithin is mixed with melted fat in a separate container and added to the sugar-NFDM mix at 70°C (158°F), followed by emulsification in a colloidal mill. The reconstituted milk should be used within 24 hr.

Fudge, caramel, and toffee consist of sugars, milk solids, vegetable fat, emulsifier, salt, and flavorings (Table 6-3). Caramel tends to contain more milk solids than fudge or toffee. Fluid milk is not the ingredient of choice because of the necessity of evaporating large volumes of water during confection preparation. Accordingly, concentrates of milk solids are favored for their handling ease and superior functional performance. Condensed milk (30% solids) and butter produce the desired flavor in heavy-bodied caramel and fudge. Flavor and body are parameters of a good quality product (4). Evaporated milk gives a darker color, more intense flavor, and a smooth, heavy body to the confections. In comparison with whole milk concentrates, NFDM does not produce the same quality of confection.

The degree of heat applied during the manufacture of milk concentrates affects the physical state of the milk protein, which in turn affects the body and texture of the confection. For example, excessive denaturation of protein results in poor texture in caramels.

The addition of lipolyzed butteroil produces creamy and buttery flavors, especially in butterscotch and caramel candy products. Milk proteins are used in caramel and fudge to impart proper body, flavor, and color. Caseins cause whitening and enhance moisture retention in toffee and caramel. Dry whey and reduced-mineral whey solids offer an economical partial replacement of milk solids in caramels.

Whey and lactose concentrates that are lactase-treated to produce protein-containing syrups with a sweetness level of 42 dextrose equivalent are suitable in the production of caramels and toffees. Lactose-hydrolyzed products may be useful in reducing lactose and controlling grainy texture problems in toffee. When used in confections at concentrations higher than 5%, lactose crystallizes during shelf life of the product. Lactose and sucrose together tend to alter their crystalline forms to yield a softer, smoother texture in confections.

MANUFACTURING CONSIDERATIONS

Caramel and Toffees. The manufacture of caramel involves emulsifying fats in milk or milk syrup and blending them with sweeteners, followed by boiling at 118°C (245°F) in a batch or continuous mode. The objective of cooking is to reduce moisture to 5–12%. Soft caramel contains 9–10% moisture and is obtained by boiling at 118–120°C (245–248°F), while hard caramel contains 5–6% moisture and is boiled at 128–131°C (262–267°F). The desired hardness in caramel is therefore a function of boiling temperature range, which reduces the moisture in the product. The texture is plastic at ambient temperature. Caramel is given a form by extrusion, or it may be deposited as sheets, ribbons, ropes, or individual patties (6). Softer caramel lends itself to coatings and to extrusion as layers.

Toffees and nut brittles are made by processes similar to that for caramel. They contain lower fat and milk solids. The syrup is cooked at 152–155°C (305–310°F).

Fudge. The manufacture of fudge is similar to that of caramel except that fondants (or creams) are incorporated in the formulation. All the ingredients except fondant and vanillin are cooked together in a kettle and brought to 60°C (140°F) with stirring. At this point, the kettle steam is shut off, and the blend is stirred at high speed for 5 min to homogenize it. Cooking is then resumed until a temperature of 113–120°C (237–248°F) and the desired moisture level are attained. The mixture is cooled to 90°C (194°F), and the fondant and other ingredients (chocolate, nuts, salt) are blended in, followed by addition of vanillin. The fudge is tempered, shaped, cooled, and wrapped.

Other confections. Truffles are made from cream, milk chocolate, and sweetened condensed milk, which may also contain liqueurs. A typical recipe uses equal weights of milk or dark chocolate and sweetened condensed milk. The condensed milk is mixed with melted chocolate in a planetary beating machine for 3–5 min, after which liqueur is added. The mixture is poured into shaped trays and allowed to set.

Nougats contain milk protein, egg white, and soy protein to facilitate whipping to a light foam (frappe), which is blended with sugars, fruits, and nuts. Dry milk may also be incorporated. Milk is important to nougat texture. Milkfat adds desirable flavor and acts as a lubricant. It also affects the texture of the candy by controlling sugar crystallization.

Butterscotch is made by melting sugar to a reddish brown color; adding water, butter, and cream; and cooking at 110°C (230°F). Its formulation consists of 41.5% granulated sugar, 37.6% water, 10.4% butter, 9.4% heavy cream, and 1% vanilla.

New Opportunities in Chocolate and Confections for Dairy-Based Ingredients

Dairy-based ingredients are a significant cost factor in the manufacture of confectionery products. Cost reduction strategies may include evaluating the contribution of dairy solids furnished by various sources within the framework of regulatory guidelines. Whey solids or fractions may provide acceptable functionality at reduced cost in certain formulations. In this regard, demineralized whey, whey protein concentrates, and lactose may offer economic opportunity. The functional attributes (emulsifying ability, foam formation, and water absorption) of whey proteins may provide a product developer with some innovative tools.

Milk chocolate has regulatory constraints, but compound coatings, caramel, fudge, toffees, and other confections may take advantage of these ingredients. Milk concentrates can be custom-manufactured for use in confectionery products to ensure reaction-derived flavor and texture attributes. In this respect, coprocessing and codrying may lead to a more economical confectionery ingredient.

Modifications in processing may improve the functionality of the dairy-based ingredient in confection manufacturing. For example, flow properties of chocolate could be improved by modifying the parameters of a spray dryer to optimize the moisture, particle size, or water absorption properties of dry milk. An exceedingly low moisture level (e.g., 1%) in dry dairy ingredients is preferred for chocolate manufacture because it produces more viscous chocolate. Thus, an ingredient manufacturer has the opportunity to balance low moisture levels and desirable flavor attributes. Furthermore, new processes may be designed to treat the powders with cocoa liquor to modify flavor.

Dairy processing techniques play a major role in the final flavor of confectionery products. A standard, consistent flavor for the ingredients helps the confectionery producer to predict the final flavor of confections. The shelf life of ingredients must be reliable. Processing conditions should ensure lipase-free ingredients to eliminate the rancid flavor defect in chocolate. Thus, milk processing technologies can be tailor-made to facilitate the standardization of current products and the development of new confectionery products.

Troubleshooting

CHOCOLATE AND CONFECTIONERY PRODUCTS		
Symptom	**Causes**	**Changes to Make**
Soapy, rancid flavor	Lipase activity	Check incoming ingredients for off-flavors.
Burnt, medicinal, fishy, or chemical taste	Oxidation of oils or fats	Minimize exposure to light and maintain cool temperatures during product storage. Check incoming ingredients for off-flavors.
Scorched or burnt taste	Overheated sugar and milk solids	Control temperatures and times during heating and cooling. Check incoming ingredients for off-flavors.
Lack of brown color and flavor notes	Insufficient Maillard browning	Increase amount of milk products. Use a slower cooking process.
	Too little caramelization	Increase sugar content. Cook at higher temperatures and/or longer cooking times.
Poor caramel texture	Too thick or poor elasticity	Decrease milk protein level.
	Too thin	Increase milk protein level to increase viscosity.
Gritty or grainy texture	Change in sugar crystal type	Adjust balance or levels of sugar types for proper crystallization.
	Large sugar crystals formed	Reduce lactose content. Check seeding and agitation during heating and cooling.
	Sugar crystals formed on product surface	Optimize storage conditions (humidity and surface temperature) to prevent migration of fat and water to surface.
White, dusty appearance	Fats not in proper phase	Check melting temperatures and cooling temperatures and times. Check amount and type of seed materials to ensure proper tempering. Increase tempering time.
	Incompatible fats	Use fats with melting points that are appropriate for the processing conditions. Avoid lauric fats in coatings. Include milkfat in formula.
	Improper storage temperatures	Store at recommended temperatures (27–30°C, 80–86°F).

References

1. Minifie, B. W. 1989. *Chocolate, Cocoa and Confectionery: Science and Technology*, 3rd ed. Van Nostrand Reinhold, New York.
2. Bouzas, J., and Brown, B. D. 1995. Interactions affecting microstructure, texture, and rheology of chocolate confectionery products. Chapter 16 in: *Ingredient Interactions: Effects on Food Quality*. A. K. Gaonkar, Ed. Marcel Dekker, Inc., New York.
3. Haylock, S. J., Towler, C., and Hewitt, S. A. 1995. Dairy component interactions in food products. Chapter 10 in: *Ingredient Interactions: Efects on Food Quality*. A. K. Gaonkar, Ed. Marcel Dekker, Inc., New York.
4. Campbell, L. B., and Pavlasek, S. J. 1987. Dairy products as ingredients in chocolate and confections. Food Technol. 41(10):78-85.
5. Kinsella, J. E. 1970. Functional chemistry of milk products in candy and chocolate manufacture. Manufacturing Confectioner 50(10):45-54.
6. Boutin, R. F. 1992. Confections. In: *Encyclopedia of Food Science and Technology*, Vol. I. Y. H. Hui, Ed. John Wiley, New York.
7. Stauffer, C. E. 1996. *Fats and Oils*. American Association of Cereal Chemists, St. Paul, MN.
8. Beckett, S. T., Ed. 1988. *Industrial Chocolate Manufacture and Use*. Van Nostrand Reinhold, New York.

Sauces, Dressings, and Dairy Desserts

Sauces and dressings include mayonnaise, salad dressing, and condiment sauces (ketchup, salsa, pasta sauces, etc.). From the standpoint of dairy ingredient use, cheese sauces, cheese dressings, and milk-based puddings are the major products of commercial importance.

Dairy-based ingredients contribute to the flavor, color, appearance, body, and texture of sauces, dressings, and puddings. In some products, such as cheese sauce or dips, dairy ingredients are the primary ingredients in the base of the product. In others, such as salad dressings, they may be used in smaller amounts for flavor or appearance.

Cheese Sauces/Dressings

Sauces derived from Cheddar and other types of cheese constitute a significant and specialized business for ingredient suppliers. The popularity of these products results from their convenience for preparing foods in food service and fast-food operations. Their main applications are in preparation of sandwiches, omelets, nachos, and pasta dishes and for use as a topping on potatoes and vegetables.

Cheese sauces are generally manufactured by *low-acid thermal processing* and have an aseptic packaging system. The manufacturing procedures are mainly in the proprietary domain of private companies. A primary innovation in cheese sauce is the ability to develop and retain desirable flavor, body, and texture when subjected to "commercially sterile" heat processing temperatures and times. A patented formulation for cheese sauce is shown in Table 7-1. Many varieties of cheese sauce are also available in powder form.

Wide variations exist in cheese sauce flavor. Besides mild and aged Cheddar, Mexican food flavors like salsa, nacho, and jalapeño have been developed and marketed.

MANUFACTURE

In the processing of cheese sauce, cheese, water, *emulsifying salts*, and other dry ingredients are incorporated with agitation to form a slurry for aseptic processing and packaging. A published report (2) describes the system, which consists of a high-shear blender, product surge tanks, balance tanks, a homogenizer, a tubular heat exchanger,

In This Chapter:

Cheese Sauces/
Dressings
  Manufacture
  Quality Evaluation

Dressings and Dips
  Dairy Salad Dressings
  Sour-Cream-Based Dips

Puddings

Troubleshooting

Low-acid thermal processing—Processing in which foods, other than alcoholic beverages, are made to have a finished equilibrium pH higher than 4.6 and water activity more than 0.85. Used with aseptic packaging.

Emulsifying salts—Salts involved in emulsification of fat and stabilization of the emulsion.

TABLE 7-1. Formulation of Cheese Sauces[a]

Ingredient	Percent
Carrageenan gum	0.06
Locust bean and guar gum	0.3
Modified waxy maize starches	6.0
Whey protein concentrate	0.2
Salt	0.8
Onion powder	0.3
Garlic powder	0.03
Water	74.0
Grated Cheddar cheese	11.8
Vegetable oil	5.5
Lactic acid	0.05
Natural cheese flavor or enzyme-modified cheese	0.6
Natural yellow coloring	0.03
Disodium phosphate	0.5
Spice and seasoning	0.2

[a] Adapted from (1).

Thermoduric bacteria— Bacteria that survive pasteurization.

a preheater, four scraped-surface heaters, a holding tube, 12 scraped-surface coolers, and an aseptic filler for No. 10 cans. To meet filling requirements of 30 cans per minute, the continuous aseptic processing system operates at about 22 gal/min. There is no sterile surge tank ahead of the filler.

Cheese sauce ingredients are mixed in the high-shear blender, and the slurry is pumped into surge tanks equipped with agitation and temperature control. The slurry is then fed into the balance tank, where its temperature is raised to 43.3°C (110°F) with a heat exchanger, using energy from the condensate from the heaters. The warm slurry is then homogenized and preheated to 76.7°C (170°F) by the tubular unit before entry into scraped-surface heaters and the holding tube. Here, the sauce is heated quickly to 138.9°C (282°F) and held for 21 sec to achieve commercial sterility. Cooling of the sterilized product to 37.8°C (100°F) is achieved quickly by running the sterilized sauce through 12 scraped-surface coolers. Flavor quality is maintained by precise time-temperature control during heating and cooling cycles. The cheese sauce is poured into appropriate containers, which are sealed aseptically (3).

The canned product is held for 14 days at the manufacturing plant for quality control clearance. Samples drawn from manufacturing runs are incubated at 37.8°C (100°F) for two weeks. Quality checks include integrity of can seams and vacuum, pH, color, standard plate count, and *thermoduric bacteria* count. Aseptically processed cheese sauces require no refrigerated storage until they are opened. After opening, the sauce should not be held longer than 10 days at 4.4°C (40°F). The sauce is pumpable and provides good portion control in a fast-food operation. Versatility of flavor and thick or thin body can be formulated. The skin that may form on the surface of cheese sauce exposed to air or steam can be avoided by covering the can with aluminum foil and stirring occasionally.

QUALITY EVALUATION

Sensory evaluation of cheese sauce is conducted in *as is* form. It is also tested as a topping on potatoes or nachos or in a macaroni-cheese product. Occasionally, its flavor performance is tested in cheese sandwiches. Attributes of appearance considered desirable include uniformity of color, shiny appearance, and smooth texture. The flavor should be fresh and clean with no detectable canned, bitter, salty, or sweet flavor notes. The texture should give no perception of graininess, lumpiness, or gumminess.

Other defects in body and texture occasionally encountered are low viscosity ("watery") and syneresis. These are attributable to ingredients and/or improper processing controls. Modified starch contributes to the body and texture of cheese sauce. However, factors such as hydration, temperature, and the presence of starch-degrading enzymes and substances can affect the development of the desired body.

Dressings and Dips

DAIRY SALAD DRESSINGS

Spoonable and pourable dairy dressings displaying rich creamy mouthfeel are based on buttermilk or cultured milk. Regular salad dressings contain 30–80% fat (4). Dairy-based dressings may contain 0–12% fat. They need to be refrigerated in storage, distribution, and use. Flavors include bleu (blue) cheese, creamy thousand island, avocado, creamy garlic, creamy Italian, and creamy French. Dairy-based ingredients are also used to contribute flavor to salad dressings packaged in powder form.

Production of the dairy base involves formulation of a base preparation similar to that of cultured buttermilk. Stabilizers and emulsifiers are added to the cultured base, followed by pasteurization and homogenization. The base is then blended with salad dressing bases to obtain the salad dressing.

Figure 7-1 shows a process flow sheet for dairy-based salad dressing manufacture. For detailed information on salad dressings, including composition, oil characteristics, and manufacturing procedures, the reader is referred to *Fats and Oils* (4), another publication in this handbook series.

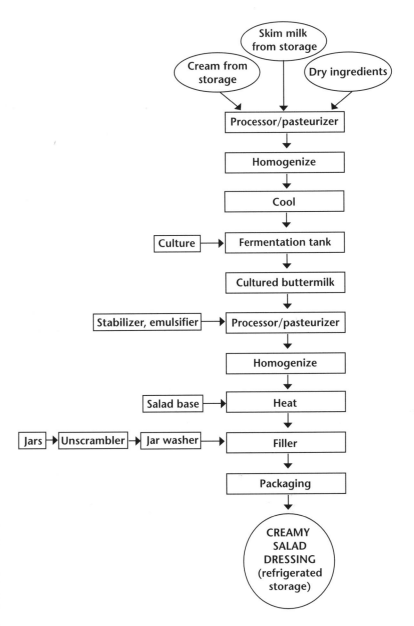

Fig. 7-1. Flow diagram for manufacture of a creamy salad dressing.

SOUR-CREAM-BASED DIPS

Many dips currently available are formulated with a sour cream base and added flavors that range from savory to sweet types. Such a dip is used with salty snacks, raw vegetables, or fruit. It must be formulated so that a chip or fruit piece can be dipped into it; it must be soft enough to be scooped with a chip without breaking the chip and must have sufficient body to stay on the chip or fruit piece. An outline of the manufacture of a 1,000 gallon batch of sour cream follows. The ingredients are skim milk (4,245 lb), 40% cream (4,057 lb), stabilizer (37 lb), and rennet (5 ml).

Stabilizer is blended into ~10% of the skim milk through a powder funnel. The cream is added and then the remainder of the skim milk to flush the line. After heating at 87.8°C (190°F) for 3–5 min, the mixture is homogenized at 71.1°C (160°F) and 2,500 psi in a single stage. It is cooled to 22.2°C (72°F) and pumped to a cone vat. At this point, 2% bulk starter (the same as for buttermilk) and diluted rennet (200 ml) are added. The mixture is mixed for 20 min and held at 22.2°C (72°F) until a pH of 4.5 or a titratable acidity of 0.75% is reached. It is then subjected to agitation to break the curd and is cooled to 18.3°C (65°F).

The product is packaged at 18.3°C (65°F) using a positive-drive pump fitted with a sour cream valve for smooth texture. It is held at 1.6°C (35°F) for 24–72 hr before being shipped (5). Figure 7-2 shows a flow diagram for sour cream and dip manufacture.

The finished sour cream has a pH of 4.5–4.6 with a heavy, smooth body, no off-flavors, and a mild acetyl buttery flavor. It must show no wheying off (syneresis), should weigh 8.357 lb/gal, and should have a shelf life of 21–30 days at 7.2°C (45°F). Microbiological standards call for <5 colony-forming units (CFU) of contaminants per gram, with coliform, psychrotrophic organisms, and yeast and mold each at <1 CFU/g.

In the manufacture of extended-life dips, sour cream is blended with nonfat dry milk and stabilizer, then pasteurized and homogenized (Fig. 7-2). Seasonings are blended in at 80°C (176°F), and the mixture is hot packed in sealed containers, producing a sour cream dip with a refrigerator shelf life of three to four months.

Puddings

Milk-based puddings have been consumed around the world for a long time. Originally, they were prepared in the kitchen from raw ingredients. Dry mixes containing starch, flavor, and color provided more convenience at home. The advent of shelf-stable ready-to-eat puddings in single-service containers has given the product both convenience and portability.

Milk contributes nutritional attributes—protein, minerals (especially calcium), and vitamins—to puddings and provides a traditional and natural base.

Milk-based pudding products are made from skim milk and fall under the low-fat or nonfat designation of the Nutritional Labeling and Education Act. Traditionally, pudding was produced from milk and corn starch using batch processing in vats. Corn starch gave a typical consistency and texture to the pudding and a "starchy" eating quality. Modified waxy maize starch produces a smooth texture, imparts stability to the product even at low temperature, and provides temperature and shear resistance. Modified tapioca improves spoonability. Puddings can be formulated and processed for nonsterile or sterile-aseptic packaging systems. The sterile product with long shelf life is distributed and marketed under ambient storage

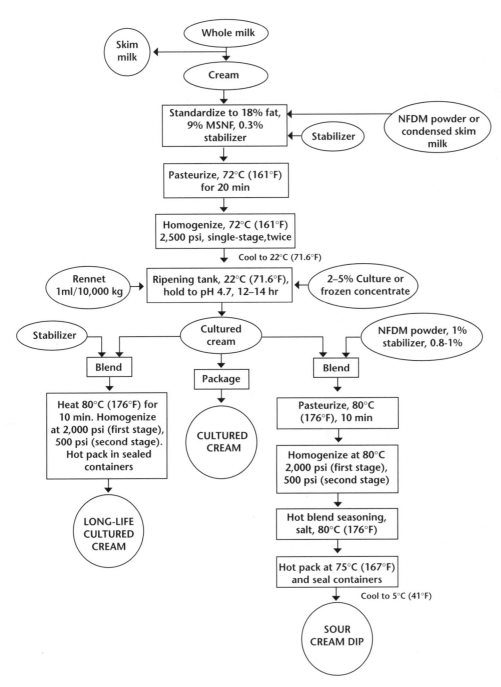

Fig. 7-2. Flow diagram for manufacture of sour cream-based dips. MSNF = milk solids-not-fat; NFDM = nonfat dry milk; long-life cultured cream = sour cream with a shelf life of three to four months at 5°C (41°F). From: Other Fermented Dairy Products, by R. C. Chandan and K. M. Shahani, in: *Biotechnology,* 2nd ed., Vol. 9, G. Reed and T. W. Nagodawithana, Eds. ©1995 by VCH Publishers. Used by permission of John Wiley & Sons, Inc.

TABLE 7-2. Formulation (lb) for a 1,000-Pound Batch of Dairy Pudding

Ingredient	Vanilla	Light Chocolate	Dark Chocolate
Whole milk	750	732	726
Cream (40% fat)	25	25	25
Nonfat dry milk, low heat	18	13	13
Sucrose	148	160	160
Modified starches (blend)	58	57	51
Flavor and color	As needed	As needed	As needed
Cocoa	...	14	25

TABLE 7-3. Formulation for Sterilized Pudding

Component	Vanilla	Chocolate
	Percent in Pudding	
Sucrose	14.77	14.77
Nonfat dry milk, low heat	5.18	5.18
Corn syrup solids	2.46	2.46
Guar gum	0.12	0.12
Carrageenan	0.05	0.05
Cocoa	...	2.92
Salt	0.09	0.09
Modified starches	3.12	3.00
Emulsifiers	0.13	0.13
Vegetable fat	6.61	6.60
Water	67.47	64.68
Color	As needed	As needed
Flavor	As needed	As needed
Total	100	100

conditions, while the short-shelf-life non-sterile pudding must be refrigerated at all times.

Puddings made with corn starch yield an unsightly mixture of gritty curd and whey on freezing and thawing. Also, after refrigerated storage for one to two weeks, the starch gel shows shrinkage and there is a tendency for a milky liquid to separate out (6). When modified starches are used, nonsterile puddings have built-in freeze-thaw stability, which allows extended shelf life when kept in frozen storage.

Nonsterile dairy puddings may be manufactured by the hot-pack procedure described below. Table 7-2 shows the formulation for a large batch. Milkfat in the pudding can be varied using milk or skim milk blends, or vegetable fat can be substituted, if desired. The typical composition of vanilla pudding involves standardization to 3.5% milkfat, 8.25% milkfat not solids, 14.75% sucrose, and 5.8% modified starch (a tapioca-starch blend). Chocolate pudding may contain 3.5% milkfat, 7.5% milkfat not solids, 16% sucrose, and either 1.4% cocoa (for light chocolate pudding) or 2.5% cocoa (for dark chocolate pudding).

Ingredients are blended by adding the dry ingredients to the mixture of milk and cream in the mix vessel. Flavor and boiled color solution are added during heating to 65.6°C (150°F). The mix is pumped to the processor, where it is heated to no more than 67.2°C (155°F) and held for 30 min to meet pasteurization requirements.

The pasteurized product is then pumped to the pudding thermutator. The product temperature is raised to 90.5–93.3°C (195–200°F) with steam pressure and kept at 83.8°C (165°F) as the product is moved to the hopper of the packaging assembly. After pudding cups are filled, they are inverted in the wire cases with adequate interspace for proper cooling and are held at room temperature for 10–15 min for pasteurization of the interior of the cup. Cups are then cooled to 10°C (50°F) by being transferred into a freezer at –34°C (–30°F) and fanned extensively for approximately 1 hr. The product is transferred to the cooler and stored overnight at –15°C (5°F) before being put in shipping cases.

Tests are run for flavor, texture, starch granule size, fat, total solids, and sugar. In addition, microbiological tests must show a standard plate count of <100 CFU/g and yeast and mold, coliform, and psychrotrophic counts of <1 CFU/g. The product should have a shelf life of 30 days at 7.2°C (45°F).

Sterilized, aseptic pudding is subjected to a more rigorous heat treatment (3). A suggested formula is shown in Table 7-3.

Dry ingredients are blended and then slowly added to water at 26.7°C (80°F) with agitation. The mixture is heated to 48.9°C (120°F), and a mixture of vegetable oil and emulsifier at 48.9°C (120°F) is added to the aqueous blend. Color and flavor are added, and the mixture is homogenized at 48.9°C (120°F). Cooking is at 121.1–148.9°C (250–300°F) through sterilization equipment. After a holding period of 2–3 sec, the product is cooled to 26.7°C (80°F) and held in aseptic storage. It is packed in an aseptic packaging system at ambient temperature.

Troubleshooting

CHEESE SAUCE		
Symptom	Causes	Changes to Make
Poor texture	Viscosity too thick	Decrease level of thickening agents and check thickening agent selections. Reduce heat exposure during processing. Minimize evaporation losses during processing.
	Viscosity too thin	Increase level of thickening agents and check thickening agent selections. Increase heat exposure during processing.
	Lumpiness or gumminess	Decrease level of thickening agents or whey protein concentrate. Increase blending time for uniform dispersion and hydration of thickening agents.
	Graininess	Ensure uniform blending. Raise pH to avoid protein precipitation.
	Grittiness or sandiness	Reduce level of salt in formulation. Reduce cooking to decrease extent of protein denaturation. Decrease lactose content to control crystallization.
Oiling off	Emulsion breakdown	Increase levels of emulsifiers or emulsifying salts. Use less heat. Increase blending time to disperse ingredients more uniformly.
Syneresis, weeping	Colloidal system breakdown	Increase stabilizer level and check stabilizer selection. Use more casein or cheese or less whey to bind water. Use less heat to decrease extent of protein denaturation. Ensure proper blending to hydrate stabilizer. Check mineral balance.
Oil streaking	Emulsion breakdown	Increase levels of emulsifiers or emulsifying salts, use less heat, or increase blending action to disperse ingredients more uniformly.

CHEESE SAUCE, continued		
Symptom	Causes	Changes to Make
Color streaking	Uneven color distribution	Increase blending to disperse ingredients more uniformly.
	Poor color solubility	Select color with solubility properties to match the system.
Skin formation	Air exposure and evaporation	Stir and cover during usage and cover during storage.
Rancid flavor	Lipase activity	Check incoming ingredients for off-flavors. Avoid mixing raw milk with pasteurized milk product.
Oxidized flavor	Oxidation	Minimize exposure to light and maintain cool temperatures during product storage. Check incoming ingredients for off-flavors.
Bitter flavor	Protein breakdown to bitter peptides	Use milder varieties or less aged cheeses. Check incoming ingredients for bitter notes. Check age of cheese sauce.
Improper flavor intensity	Flavor too strong	Use mild-flavored or less aged cheeses. Decrease level of flavoring agents. Decrease amount of enzyme-modified cheeses in blend.
	Too little flavor	Use strong-flavored or more aged cheeses. Add more flavoring ingredients. Use enzyme-modified cheeses in blend.

SALAD DRESSINGS		
Symptom	Cause	Changes to Make
Poor texture	Viscosity either too thick or too thin	Adjust level of stabilizers or solids content. Check enzymatic activity in added ingredients. Increase pH if too thick.
	Grittiness or sandiness	Reduce level of salt in formulation. Increase pH to decrease extent of protein denaturation. Decrease lactose content to control crystallization. Ensure adequate hydration of ingredients.
Separation	Breakdown of colloidal system	Increase stabilizer level and check stabilizer selection. Ensure proper blending to hydrate stabilizers. Check order of ingredient addition.
Rancid flavor	Lipase activity	Check incoming ingredients for off-flavors. Avoid mixing raw milk with pasteurized milk products.
Oxidized flavor	Oxidation	Minimize product exposure to light and maintain cool temperatures during storage. Check incoming ingredients for off-flavors.

PUDDINGS AND DAIRY DESSERTS		
Symptom	Cause	Changes to Make
Poor texture	Viscosity too thick	Decrease level of gums, starches, or thickening agents. Check the type of gums, starches, or thickening agents. Reduce heat exposure during processing.
	Viscosity too thin	Increase level of gums, starches, or thickening agents. Check the type of gums, starches, or thickening agents. Increase heat exposure during processing to gelatinize starch to a greater degree.
	Lumpiness or gumminess	Decrease level of starch. Increase blending for uniform dispersion and hydration of starch. Mix during cooking to ensure even heating.
	Graininess	Ensure uniform blending. Raise pH to avoid protein precipitation.
Syneresis, weeping	Colloidal system breakdown	Increase level of stabilizer or starch and check selection of stabilizer or starch. Use less heat to decrease extent of protein denaturation. Ensure proper blending to hydrate stabilizer/starch.
Separation	Emulsion or colloidal system breakdown	Decrease heat treatment. Increase level of emulsifiers or starches. Increase mixing action during processing. Check for enzyme activity.
Flavor-related problems	Too little flavor	Decrease starch level or change type of starch. Increase level of flavoring.
	Scorched flavor	Decrease heat treatment to prevent scorching and reduce residual burn on.

References

1. Spanier, H. C. 1986. Cheese sauce. U.S. patent 4,568,555, Feb. 4, 1986; assigned to Nabisco Brand, Inc., Parsippany, NJ.

2. Swientek, R. J. 1987. Co-packer gears up for low-acid, aseptic-fills cheese sauces and puddings in No. 10 cans. Food Processing, U.S.A. 48(3):120-121.

3. Chambers, J. C., and Nelson, P. E., Eds. 1993. *Principles of Aseptic Processing and Packaging, 2nd ed.* The Food Processors Institute, Washington, DC.

4. Stauffer, C. E. 1996. *Fats and Oils.* American Association of Cereal Chemists, St. Paul, MN.

5. Chandan, R. C., and Shahani, K. M. 1995. Chapter 10 in: *Biotechnology*, Vol. 9, 2nd ed. G. Reed and T. W. Nagodawithana, Eds. VCH Publishers, Weinheim, Germany.

6. Emmons, D. B., Larmond, E., and Beckett, D. C. 1970. Milk-based puddings and custards could open profitable markets for dairies. Modern Dairy 49(3):17-20.

Snack Foods, Meats, and Other Applications

Snack Foods

Snacks include a wide array of foods—cookies, crackers, candy, pizza, cakes, pies, potato chips, tortilla chips, corn puffs and curls, nuts, pretzels, popcorn, granola bars, dried fruit, fruit leather products, hot dogs and other meat snacks, yogurt, cheese products, and dough-enrobed fried vegetables, cheeses, etc. Their flavor may be savory or sweet. Dairy ingredients are used in several snack items.

The primary application of dairy-based products in snack foods is for seasonings, because they enhance the flavor and appearance of the snacks. Cheese powder and dairy-based preparations of many formulations are used as seasonings. They constitute significant sales to the snack food industry.

COLLETS

Collets are puffed second-generation snacks, e.g., cheese curls or cheese balls. To produce an extruded corn-based snack, corn meal is fed into an extruder at a constant rate along with a small quantity of water. The mixture is subjected to heat and pressure in the barrel as the extruder auger transports the meal through the extruder. Water in the corn meal dough is turned into vapor due to heat generated by friction, creating high pressure in the system. As the stream is forced between two rotating heads, a sudden pressure release results in the formation of a rope with puffed texture. The rope is cut into appropriate lengths with a rotating knife. The snack shape may be cylindrical or ball-like. After extrusion, the moisture content of collets is 8–10% (wet basis). The extruded collets are then fried in vegetable oil or baked (1,2).

Before being coated with seasoning, the collets undergo additional drying at 148.9°C (300°F) for 4–6 min, which lowers moisture to 1–2% and imparts crispness. Collets are sieved off from the fines at this stage to avoid retention and accumulation of fines in the finished product.

The extruded collets are then coated with cheese seasonings. Cheddar cheese is the most popular seasoning. Coating procedures consist of: 1) dry application, in which the extruded product is

Collets—Puffed second-generation snacks obtained by extruding corn meal.

TABLE 8-1. Cheese Coating Composition—Ranges Used in Three Formulations[a]

Ingredient	Range, %
Collette	57.8–66.0
Partially hydrogenated oil	24.0–30.0
Cheddar cheese powder	5.0–9.3
Acid whey powder	3.5
Cheese flavor	0.7–3.5
Salt	0.2–3.0
Total	100.0

[a] Adapted from (1).

Masa—Steeped and cooked corn dough (or flour) used in the manufacture of corn snacks.

sprayed with oil or a gum system, followed by dusting with dry seasonings, or 2) a slurry of flavors, salt, and spices in an oil- or water-based system, which is sprayed onto the extruded product.

The typical composition of the cheese slurry is given in Table 8-1. The slurry is made in a water-jacketed kettle equipped with mixers. Hot water at constant temperature is circulated in the jacket. Using a recirculating pump, the ingredients are slurried to homogeneous composition. The temperature of the slurry is maintained at 48–54.4°C (120–130°F) and must not exceed 62.7°C (145°F). The amount of cheese on the snack is determined by the procedure given by the Snack Food Association's Corn Quality Assurance Manual (1). Final product specifications for a fried extruded snack are 1.4% moisture, 2.2% salt, 36% fat, and 7.2% cheese (1).

CHIP PRODUCTION

For potato chips, washed potatoes are destoned and peeled. They are cut into manageable pieces, sliced, and rewashed. The slices are fried at 182–193°C (360–380°F) until the moisture is <1.5%. Chips are then salted, or dusted with seasonings, and packaged. Dry or liquid seasonings are typically applied in the final steps.

The process for tortilla chips is shown in Figure 8-1. Again, seasonings are added in the final stages.

The process for fried corn chips is similar to that for tortilla chips. A softer corn is produced by prolonged cooking in additional water. The resulting *masa* contains 50–52% water. The texture of this masa is coarser than that of tortilla masa. The masa is extruded, cut into pieces, and fried. Corn chips contain 34–40% oil, whereas tortilla chips contain 22–26%. Corn chips have a little more salt (1.5% vs. 1.0%), and for both, the moisture content should be 1.0–1.2% (1).

CHEESE SEASONINGS

Dairy- and cheese-based seasonings contribute desir-

Box 8-1. Specifications for a Cheese Powder

Composition: Cheddar cheese solids, dry whey, edible oil, enzyme-modified cheese, maltodextrins, sodium chloride, blue cheese, disodium phosphate, citric acid, color

Chemical-physical parameters

Moisture	3% maximum
Fat	30–35% or variable
Salt	6–8%
Color	25–30 Agtron units (green filter)
pH	5.0–5.5

Particle size distribution specified for adhesion

Passing 78–80% through U.S. No. 100
Retained on U.S. No. 40, 2% maximum
Retained on U.S. No. 20, 1% maximum

Microbiological standards

Standard plate count	<50,000 CFU[a]/g
Coliform	<50 CFU/g
Escherichia coli, Salmonella,	
Staphylococcus aureus, Listeria	Negative
Yeast and mold count	<50 CFU

[a] Colony-forming units.

able flavor to many snack items. Sour cream and cheese flavors are popular in crackers, potato chips, and extruded/puffed snack items.

Formulation. Cheese powders for application in snack production are normally developed specifically to satisfy the needs of snack manufacturers. The custom design includes a spectrum of flavor notes. A typical specification for cheese powder is given in Box 8-1.

Consumer choice of a snack is driven by its taste (3). If a mild cheese flavor is desired, Monterey Jack or mozzarella cheese is used in the formulation. A sharper cheese flavor is achieved by the use of Romano, blue (bleu), Parmesan, and Cheddar cheeses or by incorporating the enzyme-modified cheese concentrates of these cheeses. The cheese content of the powder determines the intensity of the flavor.

Cheese powder may also be supplemented with stronger spice flavors to complement the flavor and appearance of the base. A typical nacho seasoning formulation is given in Table 8-2. Certain herbs and spices lend themselves to being codried with the cheese slurry, but whenever there is loss or change in the character of the seasonings during drying, it is more appropriate to dry blend them with cheese powder. Development of snack seasoning takes into account the rate of application as well as the mode of incorporation into the snack. The suggested usage level of the nacho seasoning is 15% to produce cheese curls. Other usage levels are given in Table 8-3.

Because cheese is an expensive component of cheese powder, it is used sparingly. The original dried cheese blends were essentially dehydrated cheese. The most common process involved blending cheeses of desired varieties and flavor with water, color, and emulsifying salts to obtain a flavor concentrate (4). The concentrate was then spray dried by atomizing it into a hot air stream, and the powder was collected from the air stream. As the market for cheese-flavored snacks developed further, cheese solids were replaced with flavor concentrates in the form of enzyme-modified cheese or butter along with sweet or acid whey solids, buttermilk or skim milk solids, or hydrogenated vegetable oil to reduce costs. Certain nondairy ingredients (e.g., maltodextrins, vegetable fat) are included now to effect more cost

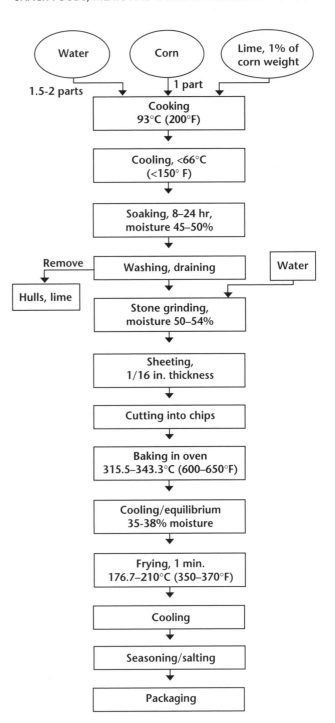

Fig. 8-1. Flow diagram for tortilla chip manufacture. (Adapted from 1,2)

TABLE 8-2. Composition of Nacho Seasoning[a]

Ingredient	Percent
Romano cheese	35.00
Parmesan cheese	10.00
Cheddar cheese	5.00
Salt, flour, or mixture	18.93
Maltodextrin	17.57
Tomato powder	5.00
Monosodium glutamate (fine)	3.00
Onion powder	1.50
Citric acid (fine crystal)	1.00
Mustard	1.00
Garlic powder	0.50
Caramel powder	0.50
FD&C yellow No. 5 lake 41% dye	0.50
FD&C yellow No. 6 lake 41% dye	0.20
Red pepper ground	0.30
Total	100.00

[a] Adapted from (5).

savings. Flavor enhancers like monosodium glutamate and autolyzed yeast extract may be used as ingredients of cheese powder. Corn syrup solids (or maltodextrins) are used as carriers and to impart flowability attributes. Table 8-4 shows the function of various ingredients commonly used in cheese powders.

Frequently, salt is added to the seasoning to provide flavor and to act as a carrier of other flavors. It also functions as a flavor enhancer and modifier. Most salty snacks contain about 2% sodium chloride. For topical application, the salt size and form are important to the dispersibility of the seasoning on the surface of the snack. Larger salt crystals dissolve slowly when the snack is eaten and provide a sharper lingering salt taste as compared with small crystals. Appearance and adhesion properties are also influenced by the salt crystal size.

To guard against deterioration of the oil in fried snacks, only compatible flavorings are used in the seasonings. Oil-soluble flavors are preferred over water-soluble flavors for fried products. Seasoning applied to the surface of an unbaked snack cracker must be resistant to heat to avoid heat-induced interactions and volatilization.

Processing considerations. Some coating processes involve slurrying the powder and spraying the slurry onto the snack. Because the slurry is drawn from a holding tank, bulk density, particle size and shape, and the sedimentation rate of the cheese powder are important factors in avoiding plug-ups of spray nozzles as well as in assuring uniformity of coating on the snacks. Spray equipment should be appropriately designed to optimize the surface adhesion of the seasoning.

The mesh size of the seasoning powder should be compatible with the oil content of the snack. For potato chips, with an oil con-

TABLE 8-3. Typical Usage Level of Cheese Seasonings in Snacks[a]

Type of Snack	Mode and Amount Applied		
	Topical	Oil Slurry (55–60% oil, 40–45% cheese powder)	Incorporated into Flour Dough
Popcorn	6–12% weight of popcorn	22–29% weight of popcorn	⋯
Potato chips	5–8% weight of chips	⋯	⋯
Corn collets/balls	6–12% weight of collets	22–29% weight of collets	Up to 10% flour weight
Cracker (dough)	⋯	⋯	8–10% flour weight
Cracker (filling)	⋯	⋯	30–35% weight of filling

[a] Adapted from (1).

tent of 36–38% by weight, a relatively coarse particle size of 40–100 mesh is adequate for optimum adhesion. Tortilla chips with less oil (18–22%) require a much finer particle size. For low-fat snacks, it may be necessary to form a surface film to facilitate adhesion of seasoning. The film is created by spraying a solution of gum acacia, corn syrup solids, or maltodextrins onto the surface of the snack and drying at 121.1°C (250°F) in convection heating equipment. In some

TABLE 8-4. Ingredients and Their Roles in Cheese Powder Formulation

Ingredient	Function
Cheese, various types	Flavor quality/intensity
Salt	Taste/adhesion of seasoning to snack
Citrates/phosphates	Cheese emulsifiers/slurrying agents
Enzyme-modified cheese/butter	Flavor booster/rounded flavor profile
Corn syrup solids	Sweet rounded flavor/carrier/flow agent
Maltodextrins	Carrier, no-flavor impact/flow agent
Whey/whey products/lactose	Filler/flavor carriers
Color	Appearance improvement
Vegetable oil	Adhesion/filler
Monosodium glutamate/autolyzed yeast extract	Flavor booster
Citric acid/sodium citrate	pH adjustment
Sodium silicoaluminate/silica dioxide/magnesium carbonate	Flow agents (anticaking)
Propylene glycol/wetting agent	Antidust agent

cases, a superior flavor profile can be achieved by using cheese powders in the formulation of the dough, followed by topical application. For cheese crackers, the cheese preparation is incorporated into the dough.

The color of cheese powders ranges from white to deep orange, depending on regional or manufacturer's preferences. The color of the seasoning imparts a desirable color to the snack indicative of the flavor being conveyed to the consumer. Colors may be of vegetable origin (turmeric, paprika, beet, annatto, b-carotene) or of synthetic origin (FD&C-permitted colors). They should make the product bright and exhibit stability during its shelf life.

The nutritional contribution of cheese powders to snacks is generally not very significant because the application rate is low (10%). Quality control tests for snacks involve sensory evaluation and determination of moisture, oil, salt, and the amount of cheese coating. Other tests frequently performed are color and bulk density.

Meat Products

During manufacture of processed meat products, a key objective is the formation of a stable, heat-induced gel (emulsion) where water and fat are bound to yield a palatable and attractive product. A meat emulsion is a multiphase system. The *continuous phase* is aqueous and contains soluble salts and proteins. Dispersed in this continuous phase are solid fat particles, muscle fibers, and connec-

Continuous phase—A liquid phase of an emulsion in which another immiscible liquid is dispersed (the dispersed phase).

Comminution—Reduction to minute particles.

Reformed hams—Hams processed by cooking pieces together to form a single piece.

Pumped hams—Hams with added water containing curing salts.

tive tissue. Emulsion stability is achieved by dispersing the fat as fine particles and creating an interfacial protein membrane at the fat-water interface.

To achieve fine particles, the fat, muscle, and connective tissues are ground. This *comminuted* meat consists of individual meat particles interspersed in a water medium containing protein capable of displaying adhesion properties. Accordingly, proteins perform an important structural and functional role. The functional attributes of proteins in contributing to the quality of meat systems are water binding (protein-water interaction), fat binding (protein-lipid interaction), meat binding (protein-protein interaction), viscosity, solubility, gelation, and surface activity.

Dairy ingredients can be used as functional ingredients in the manufacture of processed meat, poultry, and seafoods. For this purpose, lactose and milk proteins are generally used.

The application of dairy ingredients is determined largely by the degree of comminution of the meat in the formula. Hams and bacon contain no comminuted meat. Sausage, bratwurst, hamburgers, meat patties, meat balls, and chicken nuggets are obtained from coarsely comminuted meat. Frankfurters, hot dogs, bologna, meat loaves, and liver sausages are derived from finely comminuted meat.

FUNCTIONAL PROPERTIES OF DAIRY-BASED INGREDIENTS

In the manufacture of *reformed hams,* lactose is used at a level of 0.5–2%. It improves water-holding capacity in *pumped ham* and other coarse or noncomminuted meats. In the finished meat product, lactose (e.g., 1–2%), along with salt (2–3%) and phosphate (0.3–0.5%), helps in extraction and swelling attributes of meat protein. Lactose increases the ionic strength of the brine mixture and masks the bitter aftertaste associated with salt and phosphate. It enhances the cured color of ham, and it improves the sliceability and yield of cooked ham, cooked sausages, and liver products (6). Lactose-containing products like whey and nonfat dry milk are functionally not as good because of their calcium content, which interferes with the binding properties of meat proteins.

Milk protein contributes to the stability of the processed meat emulsion. Milk protein or milk protein hydrolysates can be used at a level of 0.8–1.6%. Sodium caseinate competes favorably with meat proteins for absorption at the oil-water interface to stabilize fat in the emulsion. Other functional properties of sodium caseinate that are important in meat systems are that they do not gel with heat; the denatured form improves functionality; and they can contribute to the development of viscosity.

Whey proteins as such lack these attributes and are not as useful in meat emulsions. However, recently, a whey protein texturizer has been developed with enhanced water-binding and emulsifying characteristics. The whey proteins are heat treated to denature and open up the protein molecule. Gelation of the protein is induced by further heating and salt treatment. Without heating, the whey protein texturizer gels and exhibits high viscosity in solution as compared to whey protein concentrate. *Surimi* and pressed ham or bacon are application areas of the whey protein texturizer.

PROCESSING CONSIDERATIONS

Milk proteins for use in meat processing must be in hydrated form to achieve best functionality. Accordingly, the powder should be blended with chopped meat before ice or water is added. Another procedure involves preparing a slurry of one part milk protein in six parts water in a chopper and incorporating the slurry during comminution. Alternatively, a dispersion of five parts fat, one part milk protein, and five parts water may be prepared and blended with the meat emulsion at a 10–25% level.

Temperature control during meat processing is important to the stability of the meat emulsion. Normally, the temperature should not exceed 18°C. Because milk proteins have a wider temperature tolerance, milk proteins added to meat permit a wider range of temperature.

Poultry rolls and bolognas as well as chicken nuggets or patties acquire firmer texture, improvement in sliceability, and increased juiciness when sodium caseinate is used as an ingredient. A preemulsion consisting of one part sodium caseinate, five parts chicken skin, five parts chicken fat, six parts water, and two parts ice is prepared and is subsequently incorporated into the product at the 10% level.

Turkey breast production can use sodium caseinate and milk protein concentrates to increase yield 3–6%. Usage levels of milk protein range from 0.8 to 1.5%.

Seafood processing, especially tuna, can benefit in terms of juiciness and a 6–8% increase in drain weight from the use of milk protein hydrolysate. A suggested formula for water-packed tuna employs 75% steam-cooked tuna and 25% brine. The brine contains 86% water, 4% salt, and 10% milk protein hydrolysate (6).

Other Applications

INFANT FORMULA

Infant formula based on cow's milk represents simulated human milk in chemical composition and nutritive value (7,8). Cow's milk and human milk have major qualitative as well as quantitative differences in protein, fat, lactose, and mineral content. To make infant formula, the protein composition of cow's milk is modified to make it similar to that of human milk.

Conversion of cow's milk to infant formula involves the following: 1) reducing minerals, especially the sodium content, 2) reducing the protein level, 3) changing the casein-whey protein ratio from 80:20 to 20:80, 4) altering calcium-phosphorus ratio from 1.2:1 to 2.0:1, 5) increasing the lactose content, 6) increasing the vitamin level, and 7) replacing milkfat with vegetable oil to simulate the fatty acid profile of human milk (7,9).

Surimi—A processed fish product made from mild-flavored white fish that is given the shape, texture, and flavor of lobster, crab, scallops, or shrimp.

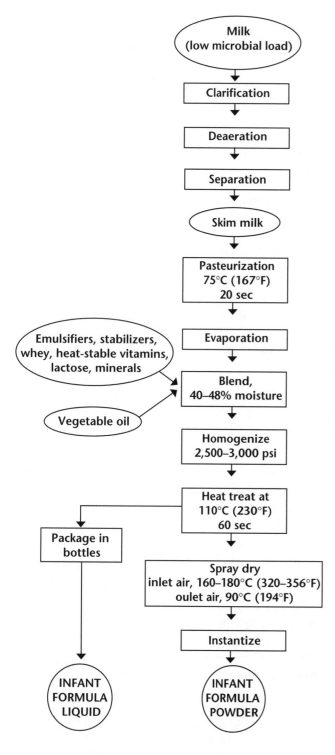

```
          Milk
  (low microbial load)
          ↓
     Clarification
          ↓
      Deaeration
          ↓
      Separation
          ↓
       Skim milk
          ↓
     Pasteurization
     75°C (167°F)
        20 sec
          ↓
      Evaporation
          ↓
Emulsifiers, stabilizers,
whey, heat-stable vitamins,  →  Blend,
lactose, minerals              40–48% moisture
                              ↑
    Vegetable oil
          ↓
      Homogenize
    2,500–3,000 psi
          ↓
     Heat treat at
   110°C (230°F)
        60 sec
       ↙        ↓
Package in      Spray dry
  bottles       inlet air, 160–180°C (320–356°F)
                oulet air, 90°C (194°F)
                    ↓
                Instantize
       ↓            ↓
   INFANT        INFANT
   FORMULA        FORMULA
   LIQUID         POWDER
```

Fig. 8-2. Typical process for infant formula manufacture. (Adapted from 7,9)

Commercial infant formulas are available in ready-to-feed, liquid concentrate, or powder forms. They are formulated from nonfat milk solids, whey (demineralized), vegetable oils, lactose, and corn syrup solids and/or sucrose.

Key steps in the manufacturing of infant formula are outlined in Figure 8-2.

Nutrient requirements in infant formulas are regulated by the FDA. Quality control procedures are generally very rigorous with regard to microbial populations, pesticide level limitations, mineral content, and vitamin concentration.

FUNCTIONAL FOODS

Foods, in general, provide essential nutrients for health maintenance. Foods that actually promote good health as evidenced by clinical trials are commonly referred to as *functional foods, designer foods, pharmafoods, biofoods,* or *nutraceuticals* (10). Such foods contain physiologically active components effective in preventing or treating disease and aid in promoting optimal health.

To realize the full potential of functional foods, consumers must be assured of their safety as well as their efficacy. Food science and technology have an obvious role in dispelling the myths of such foods by investigative research to separate fact from fiction. Regulatory authorities in most countries are confused as to how to deal with products such as these that are positioned at the food and drug interface.

From a regulatory standpoint, Japan leads the world in legislating functional foods. Japan's regulatory agency defines foods for specified health use as "processed foods containing ingredients that aid specific bodily functions in addition to being nutritious." These products must satisfy three requirements: 1) the product must be a food, not a powder or capsule, based on naturally occurring ingredients, 2) it can and should be consumed as a part of the daily diet, and 3) it provides a particular health function when ingested. Drinks constitute about 70% of the functional foods. Many functional foods are dairy-based and fortified with vitamins, fructooligosaccharides, minerals, and beneficial cultures. In addition, milk contains several constituents with known or suspected physiological roles (Table 8-5).

DAIRY BIOLOGICS

Bovine immunoglobulins are a new class of oral therapeutics made from the antibodies in colostrum of dairy cows and designed to attack infections in the gastrointestinal tract of humans. They are consumed orally and provide passive immunotherapy. Regular bovine colostrum contains antibodies against many human pathogens, as shown in Table 8-6.

To produce additional antibodies, cows are immunized against specific human pathogens during the dry period, thereby producing short-term increases in specific antibodies. The commercial process involves identification of a specific pathogen, immunization of pregnant cows, collection of four days' worth of postpartum milk, processing, and formulation for site-specific delivery. Some biologic products contain specific antibodies against enterotoxigenic *E. coli*, rotavirus, *Shigella flexneri*, cholera, *Clostridium difficile*, *Cryptosporium parvum*, *Streptococcus mutans*, *Helicobacter pylori*, and *Candida albicans*.

The advantages of the immunoglobulin *biologics* include safety, ability to target specific pathogens, and a wide spectrum of action against viruses, fungi, parasites, bacteria, and toxins.

PROBIOTICS AND PREBIOTICS

Milk is an excellent medium to generate live and active cultured products that fit into the current demand for *probiotics* (10,11). Probiotics, which can be used in animal feed or human food, are direct-fed microbial products containing abundant live and active cultures. They have been defined in the literature as live microbial feed and food supplements that beneficially affect the host by improving its intestinal microbial balance.

A *prebiotic* is a nondigestible food ingredient that beneficially affects the host by selectively stimulating the growth and/or activity of one or a limited number of bacteria in the colon, thus improving host health. For a food ingredient to be classified as a prebiotic, it must: 1) be neither hydrolyzed nor absorbed in the upper part of the gastrointestinal tract, 2) be a selective substrate for one or a limited number of beneficial bacteria inhabiting the colon, which are stimulated to grow, and 3) be able to alter the colonic flora in favor of a healthier composition to induce luminal or systemic effects that are beneficial to the host.

Fructooligosaccharides are the only products presently recognized and used as food ingredients that meet all of these criteria. Transgalactosylated disaccharides

Functional foods—Foods that have been shown by clinical trials to promote health, prevent disease, or help in the treatment of disorders. Also known as *designer foods, pharmafoods, biofoods,* or *nutraceuticals.*

Biologics—Biological products used in medicine.

Probiotics—Live microorganisms introduced into the gastrointestinal tract of humans or animals to improve the microbial balance or microbial metabolic activity.

Prebiotics—Nondigestible food ingredients that improve the host's health by selectively stimulating the growth and/or activity of one or more bacteria in the colon.

TABLE 8-5. Milk Constituents with a Possible Physiological Role

Component	Effect
Bioactive peptides	Immune system modulation, antihypertensive
Calcium	Osteoporosis, cancer prevention
β-Casein	Immunoglobulin enhancement
Glycomacropeptide	Prevention of dental caries, gingivitis
Immunoglobulins	Antibodies against diarrhea and other GI tract disturbances
Lactoferrin	Antibacterial
Lactoperoxidase	Antimicrobial
Lactose	Calcium absorption
Lysozyme	Antibacterial

Microbiota—The microscopic flora and fauna of a region.

Lactose intolerance—Maldigestion of milk sugar (lactose) by certain individuals, who experience abdominal pain, bloating, and diarrhea following consumption of milk and dairy products containing lactose.

Enteropathogen—A disease-producing microorganism in the intestinal tract.

Intestinal mucosa—The mucous membrane lining the intestine.

and soybean oligosaccharides may also fit this classification.

The use of microbes to fight microbes in controlling infections and diarrheal diseases is not new. Fermented milk products have been ingested for centuries. People now think that certain cultures can control toxin-producing bacteria in the gastrointestinal tract, thereby promoting good health and prolonging life. Healthy or balanced gut microflora are necessary for maintaining the health of the host, and the theory of probiotics suggests that one way to achieve this is to ingest exogenous bacteria and incorporate them into the colonic *microbiota*. Consumption of these organisms leads to improvement of *lactose intolerance* symptoms, improvement of gastrointestinal motility, control of intestinal disorders attributed to infections, anticarcinogenic condition, and anticholesterolemic effects. The terms *probiotics*, *lactic cultures*, and *lactic acid producing bacteria* have been loosely used to refer to these organisms as a group or individually. An effective probiotic agent would have certain criteria: 1) display resistance to digestion by enteric or pancreatic enzymes, 2) display resistance to acid and bile, and 3) prevent the adherence, establishment, replication, or activity of an *enteropathogen*. The mode of delivery of the probiotic would depend on its capacity to survive and adhere to the *intestinal mucosa* and on its potential for colonizing (establishing itself and further replicating) in the bowel.

The most commonly used and reported probiotics include two genera: *Lactobacillus (L. acidophilus, L. casei, L. bulgaricus)* and *Bifidobacterium (B. bifidum, B. longum, B. breve, B. infantis, B. animalis,* and other *Bifidobacterium* species)*. Both genera are found in normal intestinal flora at relatively low levels in healthy human adults. When ingested in doses of 10^6 or greater, both can be found in stools in concentrations of greater than 10^6 CFU per gram, as long as regular ingestion continues. *L. acidophilus* and related strains can adhere to the intestinal mucosa. Adherence is believed necessary for adequate, long-term colonization of the gut. Selection of suitable strains for efficacy is therefore critical. Potential mechanisms by which probiotics may exert their beneficial effects are: 1) competition with other microflora for nutrients, 2) production of acids inhibitory to certain enteropathogens, 3) production of bacteriocins or inhibitory metabolites, 4) immunomodulation, and 5) competition for adhesion to the intestinal mucosa.

Fermented dairy products made with traditional cultures are generally considered to lack any adverse effects. However, microbiologists and ecologists caution that introducing large numbers of viable *nontraditional* strains or strains of nonhuman origin may pose the potential for transfer of antibiotic resistance or for provoking virulence in generally benign strains.

Since the efficacy of a probiotic is directly

TABLE 8-6. Pathogens for Which Natural Antibodies Are Found in Bovine Colostrum

Candida albicans	*Serratia marcescens* HY
Cryptosporium parvum	*Staphylococcus aureus*
Escherichia coli	*Staphylococcus epidermidis*
Klebsiella pneumoniae	*Streptococcus faecalis*
Proteus vulgaris	*Streptococcus pyogenes*
Pseudomonas aeruginosa	*Streptococcus viridans*
Salmonella typhimurium	

related to the number of live and active culture cells consumed, it is important to specify the potency of the colony-forming units of the culture contained in the product. Additionally, the culture should be active in terms of growth potential.

There is extensive public information on probiotics products in Japan. Health claims for food probiotic products in Japan are primarily given as "improves intestinal microflora" or "improves intestinal condition" or the equivalent. Japanese consumers believe that bifidus-containing products in particular have a role to play in the prevention of cancer and other serious diseases.

Troubleshooting

SNACK FOODS

Symptom	Causes	Changes to Make
Flavor is too intense	Strong-flavored cheese varieties in cheese powder	Select powders from milder-flavored cheese varieties. Check specifications for blend.
	Excessive seasoning levels on base product	Decrease amount of seasoning applied. Add extender to base product. Decrease residence time in coating process. Decrease level of adhering agent.
	Age of cheese powder	Check handling, storage, and inventory procedures.
	Seasoning mixture not homogeneous	Ensure continuous mixing before application.
Too little flavor	Mild-flavored cheese varieties in cheese powder	Select powders from stronger cheese varieties. Check specifications for blend.
	Insufficient seasoning on base product	Increase amount of seasoning applied during coating process. Increase residence time in coating process. Increase application of adhering agent. Decrease seasoning particle size. Ensure uniform distribution during coating process.
Undesirable flavors	Development of rancid notes in cheese powders	Check storage conditions and handling procedures.
	Segregation of seasoning components	Ensure adequate mixing before coating process. Adjust blend of ingredients to compensate for differences in density and sedimentation rates

SNACK FOODS, continued

Symptom	Causes	Changes to Make
Color too intense	Excessive coating on base product	Decrease amount of seasoning applied. Decrease residence time in coating process. Decrease level of adhering agent.
	Seasoning mixture not uniform	Ensure continuous mixing before application.
	Improper cheese powder selection	Check seasoning before application. Use finer particle size. Select cheese powders with lighter colors.
Color too light	Insufficient coating on base product	Increase amount of seasoning applied. Increase residence time in coating process. Adjust level of adhering agent and particle size.
	Improper cheese powder selection	Use coarser particle size. Select cheese powders with more yellow color.
Uneven coating	Poor adhesion	Decrease particle size. Increase level of adhering agent. Ensure uniform distribution of adhering agent. If using oil as adherent, check oil temperature during processing.
	Insufficient coating	Increase amount of seasoning applied. Increase residence time in coating process. Increase level of adhering agent.
	Excessive liquid on surface dissolving seasoning	Adjust level of adhering agent.
	Lack of uniformity in mixture	Ensure adequate mixing before application.
	Incompatibility of adhering agent to seasoning mix	Select ingredients with suitable functional properties.
	Spray nozzle clogging	Use finer particle size for mixture. Ensure adequate mixing before spraying. Check temperature of mixture as well as temperature at application. Flush system more frequently.
	Clumping of seasoning mixture	Check water activity of powder. Check blending procedures. Ensure uniform distribution during coating process. Add anticaking agent to seasoning.

References

1. Snack Food Association. 1988. *Corn Quality Assurance Manual*. Alexandria, VA.
2. Altomare, R. E., Kettunen, D. M., and Cante, C. J. 1992. Snack food technology. In: *Encyclopedia of Food Science and Technology*, Vol. 1. Y. H. Hui, Ed. John Wiley, New York.
3. Kuntz, L. A. 1996. Seasoning secrets for salty snacks. Food Product Design 5(10): 53-66.
4. Chappell, R. A. Functionality of cheese products in snack foods. In*: Dairy Products for the Cereal Processing Industry*. J. L. Vetter, Ed. American Association of Cereal Chemists, St. Paul, MN.
5. Anonymous. 1979. Cheese flavors suited to economical end uses. Food Product Development 13(12):33-34.
6. Van den Hoven, M., and Van Valkengoed, B. 1992. Dairy ingredients: Application in meat, poultry, and seafoods. In: *Encyclopedia of Food Science and Technology*, Vol. 1. Y. H. Hui, Ed. John Wiley, New York.
7. Caric, M. 1994. *Concentrated and Dried Dairy Products*. VCH Publishers, New York.
8. National Dairy Council. 1993. *Newer Knowledge of Milk*. National Dairy Council, Rosemont, IL.
9. McDermott, R. L. 1987. Functionality of dairy ingredients in infant formula and nutritional specialty products. *Food Technology* 41(10):91-103.
10. Goldberg, I. 1994. *Functional Foods (Designer Foods, Pharmafoods, Nutraceuticals)*. Chapman & Hall, New York.
11. Goldin, B. R., and Gorbach, S. L. 1992. Probiotics for humans. In: *Probiotics—The Scientific Basis*. R. Fuller, Ed. Chapman & Hall, New York.

Nutrition and Labeling

Nutrients in Dairy Products

A nutrient is a constituent of food that, on ingestion, is used by the body for growth, maintenance, or repair of tissues. The six categories of nutrients are protein, fat, carbohydrates, vitamins, minerals, and water. Food energy is derived from three nutrients: carbohydrates (4 kilocalories per gram), protein (4 kcal/g), and fat (9 kcal/g). Although vitamins and minerals contribute no calories or energy to the body, they play important roles in regulating various body processes and in forming structural components.

Milk and milk products contribute a significant proportion of nutrients to the human diet. Dairy foods other than butter contribute 75% of the calcium, 33% of the riboflavin, 34% of the phosphorus, 21% of the protein, 18% of the magnesium, 18% of the vitamin B-12, 16% of the vitamin A, 10% of the vitamin B-6, 8% of the thiamin, and most of the vitamin D in the U.S. food supply (1,2). Nearly all milk available in supermarkets is fortified with vitamin A and D to ensure an adequate supply of these vitamins. It is interesting to note that these key nutrients are provided with only 10% of the total calories in the U.S. food supply. Milk and dairy products are recognized as important constituents of a well-balanced, nutritionally adequate diet. In this regard, they complement and supplement nutrients available from grains, legumes, vegetables, fruits, meat, seafood, and poultry.

VITAMINS AND MINERALS

In view of the importance of vitamins in infant nutrition, it is helpful to understand basic differences in cow's, human, and goat's milks. Goat's milk is considered an acceptable substitute for infants allergic to cow's milk protein. Table 9-1 compares vitamin levels in cow's and goat's milks in relation to those found in human milk.

The comparative mineral content of milk from cows and goats is given in Table 9-2.

Calcium, phosphorus, and magnesium are notable macrominerals found in dairy products, and their bioavailability is outstanding. Dairy products are the dominant source of calcium in the U.S. diet. The contribution of various dairy products to calcium in the American diet, as estimated by the USDA, is illustrated in Figure 9-1 (1).

TABLE 9-1. Vitamins and Other Constituents of Milk[a]

Nutrient	Cow's Milk	Goat's Milk	Human Milk
Vitamin A, IU/100 g	126	185	241
Thiamin, mg/100 g	0.04	0.05	0.014
Riboflavin, mg/100 g	0.16	0.14	0.04
Pantothenic acid, mg/100 g	0.314	0.31	
Niacin, mg/100 g	0.08[b]	0.28	0.18
Vitamin B-6, mg/100 g	0.04	0.05	0.01
Folic acid, μg/L	50	6	50
Vitamin B-12, μg/100 g	0.357	0.065	...
Vitamin C, mg/100 g	0.94	1.29	...

[a] Adapted from (4).
[b] 0.856 niacin equivalent.

Bioavailability—The degree to which a substance can be digested and utilized by the body in the amount and form in which it is present.

TABLE 9-2. Minerals, Salts, and Trace Elements of Milk[a]

Mineral	Range Reported, mg/100 ml	
	Cow's Milk	Goat's Milk
Sodium	35–50	38–56
Potassium	140–155	153–242
Calcium	115–125	85–198
Magnesium	11–14	10–36
Phosphorus	90–100	61–153
Chloride	95–110	121–204
Citrate	150–210	70–180
Iron	0.03–0.11	0.043–0.246
Copper	0.01–0.12	0.013–0.314
Manganese	0.003–0.037	0.005–0.020
Zinc	0.22–1.9	0.192–1.411
Iodine	0.005–0.07	0.0022–0.03197

[a] Adapted from (4).

Calcium is essential for proper bone growth. It is also involved in other body functions, e.g., in muscle contraction, blood clotting, nerve transmission, and regulation of blood pressure. The daily value for calcium used in labeling is 1 g per day. Its regular consumption, especially during growth phases, is critical in the prevention of osteoporosis. Dairy products supply highly assimilable calcium and phosphorus in an optimum ratio. Their bioavailability is further enhanced by the presence of vitamin D, lactose, and phosphoprotein (casein) in dairy products. The calcium content of a serving of selected dairy products is shown in Table 9-3.

Selenium, an essential trace mineral, is present at the site of glutathione peroxidase, which catalyzes breakdown of hydroperoxides. Selenium deficiency is correlated with abnormal cardiac function, and a fatal cardiomyopathy can be alleviated with sodium selenite supplementation in the diet. The groups most vulnerable to selenium deficiency are women of child-bearing age and young children. Breast milk or milk formula for infant feeding provides an exclusive source of selenium in the first six months of life. Cow's milk is not recommended for very young infants. The selenium content of milk is related to geographical location, and the *bioavailability* of selenium is dependent on its form and distribution in the food. Goat's and human milk contain higher levels of selenium (Table 9-4) than cow's milk.

Zinc and iodine are important trace elements and sodium and potassium are important electrolytes in milk and milk products. Sodium is needed to maintain normal water balance and osmotic equilibrium in the body. Potassium is important in the regulation of blood pressure. Dairy products are moderate sources of sodium and good sources of potassium.

PROTEIN QUALITY

The nutritional quality of a protein is a function of the biological efficiency with which its building blocks (amino acids) are absorbed. The absorption of a protein depends on its digestibility. In addition, the pattern and level of *essential amino acids* determine the quality of the protein. Protein quality may be expressed by any one of several methods, such as the *biological value* (BV), *net protein utilization* (NPU), *protein efficiency ratio* (PER), *chemical score* (6), or *protein digestibility corrected amino acid score* (PDCAAS).

The NPU of a protein is the BV multiplied by the percent digestibility. Thus, animal proteins that contain an excellent amount and profile of indispensable amino acids, as well as being readily digested, have a high NPU. Average measures of protein quality for milk and milk proteins are shown in Table 9-5. These data reveal that milk proteins are of very high quality.

The casein and whey proteins of milk products complement each other in the quantitative distribution of amino acids. Casein, which is used as a reference protein, is given an arbitrary PER value of 2.5. Whey proteins have a PER value of 3.2. Accordingly, whey protein concentrate with an equivalent quantity of protein exhibits superior protein quality as compared to nonfat dry milk.

MILKFAT

Milkfat represents about 48% of the calories of whole milk. It contains approximately 66% saturated, 30% monounsaturated, and 4% polyunsaturated fatty acids. In addition, 1 g of milkfat contains 4 mg of cholesterol, 37.7 international units (IU) (9.3 retinol equivalent [RE]) of vitamin A activity, and minor quantities of other fat-soluble vitamins.

Milkfat has been perceived negatively because of its saturated fatty acid and cholesterol content. Dietary cholesterol and saturated fat raise the cholesterol level of the blood and are therefore implicated in coronary heart disease. Consumers have become more aware of these associations, and consumption of low-fat and fat-free dairy products has accelerated in recent years. However, fat is a concentrated source of energy (9 kcal/g). It protects organs, insulates the body from environmental temperature effects, and aids in the transport of fat-soluble vitamins. Cholesterol also performs important functions in the body because it is involved in the synthesis of steroids and hormones. As fat is reduced in dairy products, the cholesterol content is progressively reduced. Nonfat milk and yogurt contain trace levels (3 mg per serving) of cholesterol. Low-fat varieties contain approximately 20 mg/8-oz serving.

Milkfat used in food products contributes the positive attribute of flavor but is subject to oxidative deterioration. The characteristic short-chain and intermediate-chain fatty acids (C4–C10) account for 8–10% of the total fatty acids. These water-soluble fatty acids are absorbed through the intestinal wall in intact form. They are then transported to the liver via the portal vein and are utilized for quick energy resembling sugar metabolism. This function is important in

Essential amino acids—Amino acids that cannot be synthesized by the body and must be obtained from food.

Biological value—The proportion of absorbed protein that is retained in the body for maintenance and/or growth. It is expressed as a percentage.

Net protein utilization—Proportion of protein intake that is retained (calculated as biological value times digestibility).

Protein efficiency ratio—Gain in body weight divided by weight of protein consumed.

Chemical score—The content of the most limiting amino acid, expressed as a percentage of the content of the same amino acid in egg protein.

PDCAAS—A method to compare protein quality based on the amino acid requirements of humans. A value of 1.0 is given to proteins meeting the essential amino acid requirements for humans.

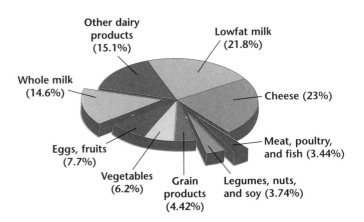

Fig. 9-1. Sources of calcium in the American food supply (adapted from 1). Sweeteners, cocoa, and coffee are included with eggs and fruits.

TABLE 9-3. Calcium Content of Some Dairy Foods[a]

Product	Calcium (mg)
Milk, fluid (8 oz)	
Whole, 3.3% fat	291
Skim	302
Cheese (1 oz)	
Cheddar	204
Creamed cottage (1/2 cup)	63
Yogurt (8 oz)	
Fruit varieties, low-fat	345
Plain, nonfat	452
Ice cream (1/2 cup, 63.8 g)	
Regular, 10% fat	88
Low-fat, soft serve (1/2 cup)	137

[a] Adapted from (5,6).

TABLE 9-4. Selenium Content of Milks[a]

Milk	Selenium content, µg/L	
	Whole milk	Skim milk
Cow's	9.6 ± 0.4	9.5 ± 0.3
Goat's	13.3 ± 0.4	13.1 ± 0.5
Human	15.2 ± 0.6	15.1 ± 0.9

[a] Adapted from (4).

early life. The emulsified nature of milkfat in milk facilitates its digestion.

The unsaturated fatty acid components of milkfat are oleic and linoleic acids and the essential fatty acids linolenic and arachidonic acids. Also, omega-3-linoleic acid, eicosapentaenoic acid, and docosahexaenoic acid are found in fairly significant levels in milkfat.

NUTRITIONAL CONCERNS

The principal carbohydrate of milk is lactose, which is exclusively found in dairy foods. In milk, it provides 30% of the total calories.

Lactose facilitates absorption of calcium, magnesium, and manganese and helps in the net retention of phosphorus. In the gastrointestinal tract, it promotes the growth of beneficial lactic acid bacteria. In addition to participating in the biosynthesis of B vitamins (biotin, riboflavin, folacin), metabolic products of lactose inhibit the growth of putrefactive bacteria, thereby modulating the ecological balance of the gut microbiota. Because of this, gastrointestinal disturbances and accompanying health-related problems may be controlled.

Lactose intolerance refers to symptoms of bloating, flatulence, abdominal pain, and diarrhea after consumption of a defined dose of lactose. The ability to tolerate lactose is dose dependent. An estimated 20 million Americans perceive themselves as lactose intolerant. However, a majority of them may tolerate a glass of milk without provoking serious symptoms (8).

Lactose intolerance symptoms can be managed by consumption of lactase enzyme, commercially available as capsules or tablets, along with milk and milk products. Lactose-hydrolyzed milk is also available; in this milk, at least 90% of the lactose has been hydrolyzed before pasteurization. The treatment imparts a slightly sweet taste to the milk because glucose and galactose (which are sweeter than lactose) are generated from lactose by lactase.

Yogurt containing live and active cultures is known to be digested without characteristic lactose-intolerance symptoms. The major reasons for tolerance of yogurt are twofold. First, yogurt fermentation reduces lactose in the product by nearly 30%. Second, the lactase enzyme contained in the live culture is released in the upper intestine. Accordingly, the lactase is made available to the lactose-intolerant individual for lactose digestion. It is interesting to note that the presence of lactase and evidence of lactose hydrolysis have been observed in the duodenal section of the gastrointestinal tract of lactose-intolerant subjects following consumption of live and active yogurt.

Lactose malabsorption, or *lactose maldigestion*, refers to insufficient digestion of lactose due to lack of adequate lactose-hydrolyzing enzyme (β-galactosidase or lactase) in the brush border membrane of the mucosa of the small intestine. It is measured by an increase in blood glucose (1.12 mmol/L) or breath hydrogen (20 ppm) following ingestion of 50 g of lactose in water. This dose is physiologically high. Clinical symptoms vary depending on the level of intact lactose reaching the large intestine, where it is fermented by colonic bacteria, producing methane and hydrogen gases. These gases are transported via the blood to the lungs and are eventually exhaled in the breath. Intact lactose can also increase the osmotic flow of water into the intestinal lumen, leading to diarrhea.

The primary type of lactose malabsorption is in human adults. With the exception of the people of Northern and Middle European descent, 70–100% of the world's adults lose 90–95% of their lactase activity after the first to third year of life, so their ability to digest lactose is impaired to varying degrees. Lactose intolerance symptoms are not necessarily observed in individuals showing lactose malabsorption. A secondary form of lactose malabsorption is induced by diseases that damage the small intestinal mucosa. Conditions leading to this defect are infections (bacterial or parasite), Crohn's disease, celiac sprue (nontropical), and protein-energy malnutrition. In rare cases, lactose malabsorption is a congenital genetic defect in newborns. These infants develop diarrhea with their first milk feeding.

MILK ALLERGY

An allergy is a hypersensitivity to a foreign body, or antigen, stimulating a specific immune response with the production of antibodies. Immunoglobulin E (IgE) antibodies are specifically related to the allergic reaction, which occurs within minutes of exposure to the offending allergen.

Infant formula based on cow's milk is adequate for the majority of the population. Cow's milk allergy can be confused with lactose malabsorption (lactase deficiency). An allergic reaction involves a true sensitivity toward a specific milk protein. Cow's milk allergy involves IgE responses, and caseins as well as β-lactoglobulins are the major allergens. Depending on the immunological definition, the rates of allergy range from 1–3 to 7–8%. Infants with cow's milk allergy are traditionally placed on soy-based formula, but 20–50% of these infants still show intolerance symptoms to the soy-based formula. Several scientists recommend use of evaporated goat's milk or milk powder in infant formula. Since the α_s-casein content of goat's milk is relatively low, it is logical that infants with sensitivity to α_s-casein should tolerate goat's

TABLE 9-5. Average Measures of Protein Quality for Milk and Milk Proteins[a]

	Quality Indicator[b]				
	BV	D	NPU	PER[c]	CS
Milk	84.5	96.9	81.6	3.09	60
Casein	79.7	96.3	72.1	2.86	58
Lactalbumin	82	97	79.5[d]	3.43	...
Nonfat dry milk	...	...	...	3.11	...

[a] Adapted from (6,7).
[b] BV = biological value, D = digestibility, NPU = net protein utilization, PER = protein efficiency ratio, CS = chemical score.
[c] PER values are often adjusted relative to casein, which may be given a value of 2.5.
[d] Calculated.

Nutrition Facts

Serving Size 1 cup (225 g)
Servings Per Container 1

Amount Per Serving

Calories 230	Calories from Fat 25

	% Daily Value*
Total Fat 2.5g	4%
Saturated Fat 1.5g	8%
Cholesterol 10mg	3%
Sodium 140mg	6%
Total Carbohydrate 43g	14%
Dietary Fiber 0g	0%
Sugars 12g	
Protein 10g	16%

Vitamin A 10%	•	Vitamin C 2%
Calcium 30%	•	Iron 0%

*Percent Daily Values are based on a 2,000 calorie diet. Your daily values may be higher or lower depending on your calorie needs:

	Calories:	2,000	2,500
Total Fat	Less than	65g	80g
Sat. Fat	Less than	20g	25g
Cholesterol	Less than	300mg	300mg
Sodium	Less than	2400mg	2400mg
Total Carbohydrate		300g	375g
Dietary Fiber		25g	30g

Calories Per Gram:
Fat 9 • Carbohydrate 4 • Protein 4

Nutrition Facts

Serving Size 1 cup (240 mL)
Servings Per Container 1

Amount Per Serving

Calories 130	Calories from Fat 45

	% Daily Value*
Total Fat 5g	8%
Saturated Fat 3g	15%
Cholesterol 20mg	7%
Sodium 125mg	5%
Total Carbohydrate 12g	4%
Dietary Fiber 0g	0%
Sugars 12g	
Protein 9g	16%

Vitamin A 10%	•	Vitamin C 2%	•	Calcium 30%
Iron 0%	•	Vitamin D 25%		

*Percent Daily Values are based on a 2,000 calorie diet. Your daily values may be higher or lower depending on your calorie needs:

	Calories:	2,000	2,500
Total Fat	Less than	65g	80g
Sat. Fat	Less than	20g	25g
Cholesterol	Less than	300mg	300mg
Sodium	Less than	2400mg	2400mg
Total Carbohydrate		300g	375g
Dietary Fiber		25g	30g

Calories Per Gram:
Fat 9 • Carbohydrate 4 • Protein 4

Fig. 9-2. Typical nutrition facts labels for fruit-flavored low-fat yogurt (left) and reduced-fat milk (right).

milk quite well. Furthermore, high heat treatment removes the sensitizing capacity of milk.

A 24-hr basic formula to provide 715 calories may consist of 12 oz of evaporated goat's milk, 24 oz of boiled water, and 3 1/2 tablespoons of corn syrup or 7 tablespoons of maltodextrins. Dilution with water is necessary to bring the potassium and chloride levels of goat's milk to desirable levels in a food designed for the first year of infant life. Infants older than six months can derive adequate nutrition from goat's milk as far as vitamin A, niacin, riboflavin, thiamin, and pantothenic acid are concerned. The deficiency of vitamins C,

TABLE 9-6. Descriptors on Food Labels

Descriptor	Basic Definition
Fat free	<0.5 g of fat per reference amount and per labeled serving
Low fat	≤3 g of fat per reference amount
Good source	≥310% DV[a] of the nutrient per reference amount
Excellent source	≥320% DV of the nutrient per reference amount
Less/reduced	≥325% reduction compared to a reference food
More	≥310% more compared to a reference food
Light	One-third fewer calories or 50% less fat than a reference food, provided the product is not already low in fat
Percent fat free	Must meet low-fat criteria
Healthy	Must be low fat; low saturated fat; contain at least 10% DV of one of vitamin A, vitamin C, calcium, iron, protein, or fiber; and have less than a maximum level of sodium and cholesterol

[a] Daily value.

D, B-6, and folic acid must be corrected by appropriate fortification. Accordingly, goat's milk is a viable dairy option to fulfill the nutritional requirements of infants and children.

Food Labeling

Under the Nutrition Labeling and Education Act (NLEA) of 1990, the FDA promulgated new labeling regulations that became effective in 1994.

Nutrients contained in all packaged foods must be declared on the label in relation to a standard reference amount (serving size) of the food. The label must declare the amounts per serving for calories, calories from fat, total fat, saturated fat, cholesterol, sodium, total carbohydrate, dietary fiber, sugar, and protein. Also, percent daily value (%DV) must be shown in relation to a 2,000-calorie-per-day diet for the above nutrients as well as for vitamins A and C, calcium, and iron. Figure 9-2 illustrates the nutrition facts labels for milk and yogurt.

The FDA regulations also address potential health claims and have permitted nine specific health claims on the relationship between diet and health. Before a health claim can be made, the product must contain (before fortification) at least 10% of one of the nutrients: vitamin A, vitamin C, calcium, iron, protein, or fiber.

Some descriptors on foods have also been defined, as shown in Table 9-6.

The *standard of identity* (SOI) of milk was changed in 1996 to make it compatible with the NLEA. Originally, when a nutrient content claim was part of the name of a food for which a SOI was established prior to enactment of the NLEA (e.g., "low-fat" milk), the food was declared exempt from the definition for the claim adopted under the NLEA. Low-fat foods, as defined by the NLEA, generally must have 3 g or less of fat per reference amount customarily consumed. However, under the old standards, use of low-fat milks allowed these foods to contain as much as 2% milkfat, which translates to 5 g of fat per 8-oz serving. Thus, 2% milk is now termed "reduced-fat milk." FDA has revoked the SOI for certain dairy products whose names include a nutrient content claim, including low-fat and skim (nonfat) milks, low-fat and nonfat yogurt, and low-fat cottage cheese. In the absence of an SOI, low-fat milk must conform to the definition of NLEA "low-fat," including the requirement that the food not be nutritionally inferior to the food for which it substitutes (i.e., whole milk).

Because vitamin A is fat soluble, the process of removing fat from milk unavoidably removes some vitamin A. As a result, lower-fat milk products are required to have at least some added vitamin A. However, because the addition of vitamins A and D to whole milk is optional, the only requirement for lower-fat milk products under the current rules is that they not be nutritionally inferior to whole milk. While the lower-fat products may be fortified to 2,000 and 400 IU of vitamins A and D per quart, respectively, they also may be fortified

Standard of identity— A legal standard, maintained by the FDA, that defines a food's minimum quality, required and permitted ingredients, and processing requirements, if any. Applies to a limited number of staple foods.

Skim milk, nonfat milk, and *fat-free milk* all refer to milk that contains no more than 0.2% milkfat. These terms may be used interchangeably.

to the levels of 500–3,000 IU of vitamin A per quart and 5–15 IU of vitamin D per quart.

The new law also provides that, under certain circumstances, safe and suitable ingredients that perform a technical effect (for example, thickeners and stabilizers) may be added to modified foods to maintain performance characteristics similar to those of the traditional food. Thus, revoking the standards should increase flexibility for manufacturers, decrease confusion about claims with conflicting meanings, and increase consumers' product choices.

References

1. Zizza, C., and Gerrior, S. 1995. The U.S. food supply provides more of most nutrients. Food Review Jan.–Apr. pp. 40-45.
2. Raper, N. R., Zizza, C., and Rourke, J. 1992. Nutrient content of the U.S. food supply, 1909–1988. *Home Economics Research Report No. 50*. U.S. Department of Agriculture, Washington, DC.
3. National Research Council. 1989. *Recommended Dietary Allowances, 10th ed.* National Academy Press, Washington, DC.
4. Chandan, R. C., Attaie, R., and Shahani, K. M. 1992. Nutritional aspects of goat milk and its products. Proceedings of the Fifth International Conference of Goats. *Recent Advances in Goat Production.* pp. 1869-1890.
5. Department of Agriculture. 1975. Composition of Foods, Dairy and Egg Products. *Agriculture Handbook No. 8-1.* USDA, Consumer and Food Economics Institute, Washington, DC.
6. Speckman, E. W. 1984. *Dairy Products for the Cereal Processing Industry.* James L. Vetter, Ed. American Association of Cereal Chemists, St. Paul, MN. pp. 55-82.
7. National Dairy Council. 1993. *Newer Knowledge of Milk and Other Fluid Dairy Products.* The Council, Rosemont, IL.
8. Miller, G. D., Jarvis, J. K., and McBean, L. D. 1995. *Handbook of Dairy Foods and Nutrition.* CRC Press, Boca Raton, FL.

Regulatory and Safety Aspects

Milk processing and marketing, as well as the manufacture of dairy products, are regulated by federal, state, and local authorities. Regulatory aspects of the manufacture of dairy products in the United States is discussed below.

Food processors should have a basic understanding of food regulations to ensure that they are in compliance. Regulations are of two basic types. A substantive regulation, which carries the same weight as a law, defines what may or may not be done. For example, labeling regulations define precisely how a label is to be prepared. The second type, an interpretive regulation, is enacted to address a law but allows some leeway in how to comply. Such regulations use the word "should." Since it takes a long time to develop regulations and pass laws, it may take years for the law to be enacted.

The dairy and food industries are regulated by various agencies of the federal (Table 10-1), state, and/or local governments.

Regulation by the Food and Drug Administration

Production of Grade A dairy products is regulated by the Milk Safety Branch of the U.S. Food and Drug Administration (FDA) (1). Product safety, labeling, packaging, and other product issues are included. In addition, other departments of the FDA are involved in product standards, labeling in general (under the Fair Packaging and Labeling Act), and matters related to overall compliance. Milk specialists represent the Milk Safety Branch's regional offices and work with the state regulatory agencies by providing scientific, technical, and inspection assistance to assure compliance with regulatory policies and procedures. Besides having a liaison with the FDA, each state's Department of Agriculture (Dairy Division or Health Department) is involved in regulating the milk production and manufacturing in its own state. Furthermore, the state enforces regulations related to the federal Grade A Pasteurized Milk Ordinance (PMO).

TABLE 10-1. Sources of Regulation for Food and Dairy Industries

Area of Business	Agency/Law
Hiring practices, unions, etc.	Department of Labor
Occupational health and safety	Occupational Health and Safety Act (OSHA)
Weights and measures	Department of Commerce
Sales to military	Department of Defense
Environment, pollution, and contaminants	Environmental Protection Agency FDA. (CFR Part 180–186)
Food adulteration	FDA. The Food, Drug and Cosmetic Act, Section 402
Sanitary operation	FDA. Good Manufacturing Practices in Manufacturing, Packing or Holding Human Food (21 CFR, Part 110) USDA. HACCP (21 CFR, Part 123)
General labeling	FDA (21 CFR, Part 101)
Nutrition	FDA. Nutritional Quality Guidelines for Food (21 CFR, Part 104)

THE PASTEURIZED MILK ORDINANCE

The PMO describes the requirements for product safety, milk hauling, sanitation, equipment, and labeling. It is very extensive and covers milk from production at the farm to shipment from the manufacturing facility. The requirements for product and packaging include the following:

- Must contain the term *Grade A*
- Must contain the identity of the plant
- Product standards of identity must be met
- Temperature: must be cooled to 7°C (45°F) or less and maintained thereat
- Must not exceed the bacterial limits specified in the PMO
- Coliforms: not to exceed 10 CFU/ml
- Phosphatase test: <1 μg/ml
- Antibiotics: no zone greater than or equal to 16 mm with the *Bacillus stearothermophilus* disc assay method or equivalent measures by other methods.

More information on the PMO is given in Table 2-2.

The National Conference of Interstate Milk Shippers (NCIMS) plays a key role in setting standards and regulations related to the PMO. Furthermore, NCIMS is involved in regulations pertaining to fabrication of single-service containers and closures for milk and milk products and in the evaluation of milk laboratories. The purpose of NCIMS is to promote the best possible milk supply for all people and to provide for unrestricted availability of milk and milk products in interstate shipment. The NCIMS operates to establish uniformity of product standards from state to state. Both producers and processors of milk are represented in the organization. They address issues related to laws and regulations governing Grade A milk sanitation (storage, handling), reciprocity between regulatory jurisdictions, and violations of reciprocity.

STANDARDS OF IDENTITY

Standards of identity (SOIs) have been established by the FDA and the U.S. Department of Agriculture (USDA) to protect consumers against fraudulent practices. Standards generally specify the types of ingredients the food must contain (mandatory ingredients), as well as those it may contain (optional ingredients). Standards also may specify minimum and maximum content requirements for valuable constituents as well as for fillers. Most dairy products with a standard of identity definition must conform to the FDA standard and the regulations published in the Code of Federal Regulations (CFR) (2). A few dairy products (e.g., butter and nonfat dry milk) are regulated by USDA grading and inspection programs. FDA has the authority to establish SOIs for foods whenever doing so will promote honesty and fair dealing in the interest of consumers. It has established more than 280 standards for staple food items, including milk, peanut butter, jams and jellies, and milk chocolate. The USDA's Food Safety and Inspection Service (USDA/FSIS) has approximately 74 standards for foods that it regulates.

Table 10-2 shows the location in the CFR of the SOIs for some dairy products.

Analytical Tests

To be sure that a product conforms to the regulatory SOI, the FDA specifies certain analytical tests to be performed on a dairy food, as shown in Table 10-3. In addition, to meet company standards of quality, safety, and cost, various analytical tests have been developed and updated by AOAC International and the American Public Health Association (3–5). In general, quality tests for milk and dairy products include analysis for chemical composition, physical attributes, microbiological quality, and sensory characteristics.

Analytical tests for milk composition are for fat, total solids, protein, lactose, ash, vitamins, and minerals.

The basic quality of milk is assessed by tests for, e.g., titratable acidity, added water, foreign materials, sanitizers, and *aflatoxins*. In addition, tests are conducted for antibiotics, pesticides, and other environmental contaminants. During milk production at the farm, certain trace materials including pesticide residue, fluorine, silicone, and heavy metals may gain entry into milk.

Aflatoxin—A highly poisonous and carcinogenic substance produced by the mold *Aspergillus flavus*. If contaminated peanuts, corn, or cottonseed are fed to cows, the milk may contain aflatoxin.

TABLE 10-2. Code of Federal Regulations: Standards of Identity for Dairy Foods[a]

Dairy Food	Location
General, definitions	21 CFR Part 130
Milk and cream, definitions	21 CFR Part 131.3
Milk	21 CFR Part 131.110
Yogurt	21 CFR Part 131.200
Low-fat yogurt	21 CFR Part 131.203
Nonfat yogurt	21 CFR Part 131.206
Cheeses and related cheese products	
Definitions	21 CFR Part 133.3
Methods of analysis	21 CFR Part 133.5
Notice to manufacturers, etc.	21 CFR Part 133.10
Asiago fresh and asiago soft cheese	21 CFR Part 133.102
Frozen desserts	
Definitions	21 CFR Part 135.3
Ice cream and frozen custard	21 CFR Part 135.110
Butter	
Butter	7CFR 58.2622
Dry buttermilk	7CFR 58.236

[a] Adapted from (2).

TABLE 10-3. Standard Methods for Analysis of Dairy Products

Test	AOAC Method[a]
Milkfat Content for Milk	16.059, 16.172
Total Solids for Milk	16.169
Milk Solids-Not-Fat Content for Milk	16.032
Titratable Acidity	16.032
Vitamin D Content	43.195–43.208
Moisture Content for Dry Dairy Products	16.192
Moisture Content for Cheese	16.233
Milkfat Content for Cheese	16.255
Phenol Equivalent (Phosphatase Test)	16.275–16.277
Fat Content for Frozen Desserts	16.287–16.059
Protein Content and PER for Frozen Desserts	16.285–286, 43.212–43.216

[a] Methods of AOAC International.

Polychlorinated biphenyls (PCBs)—Environmental contaminants that can accumulate in the food chain and gain entry into milk via animal feed. May cause a variety of harmful effects, including damage to the reproductive cycle of animals and plants.

Polybrominated biphenyls (PBBs)—A class of environmental contaminants that may be found in dairy products.

Polychlorinated biphenyls and *polybrominated biphenyls* have been detected in milk from cows that were fed contaminated feed. Antibiotics used to treat cows may gain entry into milk. Tolerance limits have been established to regulate the safety levels of these materials in the milk supply. Specialized milk tests include the Wisconsin and California mastitis tests for somatic cell counts (for samples exceeding 1 million cells per milliliter) and several other somatic cell count procedures (see Table 2-1).

Among the microbiological tests for raw and pasteurized milk, the total aerobic plate count gives a measure of total bacteria present. The coliform count is a marker of the sanitary quality of milk. The yeasts and mold count is an indicator of the spoilage tendency of low-pH products like yogurt, sour cream, and buttermilk.

Dairy testing in the industry is typically directed toward incoming milk, cream, and condensed and dry dairy ingredients to determine their suitability for use in the plant operations. In addition, freshly pasteurized milk and product mixes are tested for coliform count (violet red bile agar) as an overall index of sanitary quality. Pathogenic bacteria may gain entry in commercial dairy products as post-pasteurization contaminants or by cross contamination with raw milk. Pathogenic organisms receiving attention include *Listeria* spp., *Salmonella* spp., *Staphylococcus aureus, Yersenia enterocolytica, Escherichia coli* 0157:H7, and *Aeromonas hydrophila* due to the profound impact of their recent outbreaks. Table 10-4 shows the tests generally performed in dairy plants.

With a view to expediting results of microbiological analyses and implementing corrective actions in a timely manner, various rapid methods are being developed. Accuracy, speed, simplicity, cost, and validity are the key factors in their development.

Microbiological Safety and Sanitary Practices

A food plant is centered around the manufacture of safe and wholesome food products. Because it affects customer satisfaction, product loss, and compliance with sanitary codes, high quality in the food product is essential to survival and growth of the business. Factors such as plant conditions, manufacturing practices, housekeeping, sanitary standards, and the personal hygiene and work habits of employees and visitors assume critical importance in the control of quality, product safety, personal safety, and financial integrity.

TABLE 10-4. Typical Dairy Plant Quality Tests and Their Purpose

	Test	Purpose
Incoming material		
Milk	Direct microscopic count	Microbiological quality
	Sensory (odor, flavor)	General quality
	Titratable acidity	Freshness, handling practice
Fruits, nuts, syrups, sweeteners	Yeasts and molds	Microbial contamination
	Osmophilic yeasts	Shelf life of the product
Packaging materials	Sterility testing	Safety/shelf life of the product
Fresh products		
Fresh milk and other dairy products	Coliform	Detecting unsanitary processing or packaging conditions
	Escherichia coli or Enterobacter aerogenes	Indicator of postpasteurization contaminations
Milk and other fresh dairy products	Preincubate product in its container at 21°C (70°F)/18 hr, followed by standard plate count, crystal violet tetrazolium, gram-negative agar (plate count agar containing added pencillin)	Prediction of shelf life
Milk/cottage cheese	Preincubate milk at 7°C (45°F) for five to seven days followed by standard plate count test	Estimation of keeping quality
Yogurt	Yeast and mold count	Estimation of keeping quality
	Titratable acidity	Sensory quality
	Culture enumeration test	Culture's live and active status
Inline sampling and plant sanitation		
HTST[a]/filler or packaging machine/glycol or ice water and equipment surfaces	Preincubation followed by standard plate count and coliform count	Contamination with psychrotrophic organisms and general sanitation
Environmental air and water samples	Standard plate count and coliform count	General sanitation practices

[a] High-temperature, short-time.

GOOD MANUFACTURING PRACTICES

The food industry and the FDA have established a code of behavior called Good Manufacturing Practices (GMP). These practices are related to methods and control procedures employed in a food plant for the manufacture, processing, packaging, and storage of a food product with a view to guaranteeing its quality and safety to the consumer. The FDA has published elaborate guidelines for GMP.

- Employees working in a food plant are required to wash their hands with a sanitizing soap before beginning or returning to food handling.

- All persons working or visiting the production area must wear an authorized head covering to avoid contact of loose hair with food products.
- Sanitary precautions must be taken by employees when sneezing and coughing. An employee with infectious skin eruption, communicable disease, or other infected condition must have clearance from plant management before being allowed to handle food.
- Employees in production areas must wear clean uniforms that are changed daily, or more often if soiled for any reason. Shirts are required to be buttoned and tucked into trouser.
- Workers in the production area must not wear rings or neck or ear jewelry to keep these foreign materials from becoming mixed in with the food. Watches, pens, pencils, and loose materials must be removed before entry into the production area.
- Smoking, spitting, or chewing of tobacco is prohibited in the production as well as the storage area.
- Consumption of beverages or food is allowed exclusively in designated areas.
- People are not allowed to wear nail polish and/or perfume in production or storage areas.
- Glass containers and equipment made of glass, including glass thermometers, are not permitted in the production area.
- Good housekeeping in the production area is necessary for work efficiency and worker safety.

Postpasteurization contamination could undo product safety and quality achieved by proper pasteurization processing. All the equipment should undergo cleaning and sanitizing treatments before use. Pumps, valves, and pipes taken apart during production runs must be cleaned and resanitized before being used again.

Coliform testing usually gives a valuable clue as to the location of contamination and is of help in eliminating contaminating sources. Taking swab samples from equipment surfaces and subsequent plating is a good way to monitor cleaning and sanitizing practices. For example, a bacteria count of 1,500 colony-forming units (CFU)/cm^2 before cleaning should be progressively reduced to 60 CFU/cm^2 after cleaning with detergent, 10 CFU/cm^2 after the final rinse, and 1 CFU/cm^2 after disinfection. Bacteria double in numbers every 20–30 min. A single surviving bacterial cell can reproduce and build up to 10 million cells in 10–11 hr. Improper procedures and negligence can therefore lead to dangerous levels threatening the wholesomeness and safety of the product.

HAZARD ANALYSIS AND CRITICAL CONTROL POINTS

Hazard analysis (HA) is useful in the safe production of a food product. Defining critical control points (CCPs) helps to eliminate or control hazardous microorganisms or their toxins at specified points over the entire production sequence.

The *HACCP* principles (according to USDA guidelines) are enumerated below.

1. Assess the risks associated with growing and harvesting raw materials and ingredients and with the processing, manufacturing, distribution, marketing, preparation, and consumption of food.
2. Determine the CCPs at which hazards can be controlled.
3. Establish critical limits that must be met at each CCP.
4. Establish a procedure to monitor each CCP.
5. Establish corrective actions to be taken when a deviation occurs.
6. Establish recordkeeping systems that document the system.
7. Verify that the system is working.

HA involves ranking a food according to six hazard characteristics and assigning a risk category based on the ranking. Ranking of the hazard characteristics of a food is related to the presence of:

1. Microbiologically sensitive components
2. Processing steps to eliminate harmful organisms or their toxins
3. Postprocessing contamination with harmful organisms or toxins
4. Abusive handling and distribution
5. The potential of harmful effects due to consumer handling
6. Postpackaging heat process at the factory or during cooking at home

HA involves making a detailed diagram of the process flows and identifying the CCPs in the production sequence. The types of raw materials and the ingredients used in product manufacture are also an integral part of the analysis. Therfore, development of a HACCP system includes the following:

- Specifications for food products, including chemical, physical, and microbiological standards
- Product safety analysis, including potential microbiological problems
- Purchasing requirements for approved suppliers, whose products conform to ingredient specifications
- Adherence to GMP
- Physical systems hazard control programs, such as flow charts, CCPs, controls, monitoring of controls, etc.
- Product recall system
- Contract with packers, including standards
- Audit programs by quality assurance personnel
- Customer complaint handling program
- Responding to and recording regulatory inspections of the plant by government authorities

Hazard analysis and critical control points (HACCP)—A system of steps established by the food industry to produce safe products through anticipation and prevention of problems.

References

1. Department of Health and Human Services, Pubic Health Service, 1995. Grade "A" Pasteurized Milk Ordinance, revised. U.S. Government Printing Office, Washington, DC.
2. Food and Drug Administration. 1996. Code of Federal Regulation. U.S. Government Printing Office, Washington, DC.
3. Christen, G. L. 1993. Analyses. Pages 83-146 in: *Dairy Science and Technology Handbook*, Vol. I. Y. H. Hui, Ed. VCH Publishers, Inc. New York.
4. Richardson, G. H. 1990. Chapter 33 in: *Official Methods of Analysis*, 15th ed. AOAC International, Arlington, VA.
5. Marshall, R. T., Ed. 1992. *Standard Methods for the Examination of Dairy Products*, 16th ed. American Public Health Association, Washington, DC.

APPENDIX A.

Origin and Description of Some Off-Flavors in Milk[a]

General Cause	Off-Flavor	Sensory Description	Potential Causes
Microbiological	Malty	Burnt, caramel, Grape nuts-like flavor	Improper equipment sanitation; delayed cooling of milk, storage at 10°C (50.0°F) or above
	Acid	Tingling/peeling sensation on tongue; feeling of cleanliness after expectoration	
	Fermented or fruity	Odor similar to either sauerkraut or vinegar or to apples, pineapples, or other fruit	Raw milk stored in bulk for extended periods; old pasteurized milk
	Bitter or unclean	Persistent bitter taste after expectoration; unpleasant, musty, stale, spoiled, dirty	Temperature abuse resulting in microbial growth; certain weeds; dirty utensils
Absorbed	Feed	Aromatic and pleasant, depending on type of feed used; includes garlic and onion	Feeding cows 0.5–3 hr before milking
	Barny	Odor of a poorly maintained barn; unpleasant, persistent, and unclean aftertaste	Poor ventilation; buildup of aromatic compounds in barn
	Cowy	Cows-breath-like odor; unpleasant medicinal, chemical aftertaste	Cows with acetonemia or ketosis
Biochemical or chemical	Rancid	Soapy, bitter, unclean, blue cheese-like aroma; strong, foul, lingering aftertaste	Homogenization of raw milk; not pasteurizing milk immediately after homogenization; contamination of pasteurized milk with raw milk
	Light oxidized	Burnt, feathery, tallowy, medicinal, chemical taste	Exposure of unprotected milk (in plastic and glass containers) to UV rays from sunlight or fluorescent light
Processing	Cooked	Sulfurous, rich, caramelized, scorched, sweet	Pasteurization temperatures of 76–78°C (169–172°F); increased pasteurization times
	Foreign	May have odor or flavor not associated with milk, depending on causative agent	Improper use of chemicals, cleaners, or medications, resulting in contamination
	Flat	No odor; lacking full flavor and sweetness	Adulteration of milk with water; milk low in total solids content

[a] Adapted from Heer, A. K., Duncan, S. E., and Brochetti, D. Sensory detection of and consumer response to off-flavors in milk. Dairy, Food and Environmental Sanitation 15(8):488-493, 1995. Used with permission. ©International Association of Milk, Food and Environmental Sanitarians, Inc.

APPENDIX B.

Typical Composition of Dairy Ingredients

Dairy Ingredient	Water (%)	Fat (%)	Protein (%)	Lactose (%)	Ash (%)	Added Ingredient
Fluid milk						
Whole milk	87.4	3.5	3.5	4.9	0.7	
Skim milk	90.5	0.1	3.6	5.1	0.7	
Fluid cream						
Half & half	80.2	11.5	3.1	4.5	0.7	
Light cream	74.0	18.3	2.9	4.2	0.6	
Light whipping cream	62.9	30.5	2.5	3.6	0.5	
Heavy whipping cream	57.3	36.8	2.2	3.2	0.5	
Plastic cream	18.2	80.0	0.7	1.0	0.1	
Fat concentrates						
Butter	16.5	80.5	0.6	0.4	2.5	0–2.3% salt
Anhydrous milkfat	0.1	99.8	0.1	0.00	0.0	
Butteroil	0.3	99.6	0.1	0.00	0.0	
Condensed products						
Sweetened condensed milk	27.1	8.7	8.1	11.4	1.8	44.3% sucrose
Sweetened condensed skim milk	28.4	0.3	10.0	16.3	2.3	42.0% sucrose
Cheeses						
Cheddar	37.0	33.0	24.9	1.3	3.7	1.6% salt
Swiss	38.0	27.4	28.4	3.4	3.5	1.2% salt
Mozzarella, low moisture	47.0	23.7	21.0	2.5	3.0	1.0% salt
Cottage cheese curd	79.8	0.4	17.3	2.7	0.8	
Cottage cheese, creamed	79.0	4.0	12.5	2.7	1.4	1.0% salt
Powdered dairy products						
Dried whole milk	3.0	27.5	26.4	38.2	5.9	
Nonfat dry milk	3.5	0.8	35.9	52.3	8.0	
Dried buttermilk	3.5	5.3	34.3	50.0	8.6	
Spray dried butter	4.0	72.0				
Spray dried cream (from 20% cream)	0.6	71.1	11.1	14.7	2.4	
Whey						
Dry sweet whey	4.5	1.1	12.9	73.5	8.0	
Dry acid whey	4.3	1.0	12.3	65.5	11.0	
Whey protein concentrates						
Whey protein concentrate, 34% protein	3.5	4.0	34.5	51.0	7.0	
Whey protein concentrate, 50% protein	3.5	4.0	50.5	36.0	6.0	

continued on next page

Typical Composition of Dairy Ingredients, continued

Dairy Ingredient	Water (%)	Fat (%)	Protein (%)	Lactose (%)	Ash (%)	Added Ingredient
Whey protein concentrate, 80% protein	3.5	6.0	80.5	5.0	5.0	
Whey protein isolate, 91% protein	3.5	0.5	93.0	1.0	2.0	
Casein						
Acid casein	9.0	1.0	88.0	0.1	2.2	
Rennet casein	11.0	1.0	85.0	0.1	4.0	
Calcium caseinate	3.5	1.0	90.9	0.1	4.5	
Sodium caseinate	3.5	1.0	91.4	0.1	4.0	
Coprecipitate	4.0	1.0	89.0	1.5	4.5	
Lactose						
Food-grade lactose	0.5	0.1	0.1	99.0	0.2	
Minerals						
Dairy minerals concentrate	10.0	1.0	8.0	1.0	80.0	

APPENDIX C.

Typical Specifications for Milk Concentrates

TABLE C-1. Typical Specifications for Condensed Skim Milk

	High Solids	Medium Solids
Milkfat, %	0.4	0.3
Milk solids-not-fat, %	39.6	29.7
Total solids, %	40.0	30.0
Manufacturing procedure	Pasteurize skim milk at 72.2–73.3°C (162–164°F) for 17 sec. Condense in first effect at 74.4°C (166°F). Continue concentration at 48.9°C (120°F). Cool to 4.4°C (40°F) and standardize in holding tank.	Pasteurize skim milk at 72.2–73.3°C (162–164°F) for 17 sec. Condense in first effect at 68.3°C (155°F). Continue concentration at 43.3°C (110°F). Cool to 4.4°C (40°F) and standardize in holding tank.
Microbiological standards		
Standard plate count, CFU[a]/g	≤10,000	≤10,000
Coliform, CFU/g	≤10	≤10
Psychrotrophic count, CFU/g	≤1	≤1
Weight, lb/gal	9.888	9.448

[a] Colony-forming units.

TABLE C-2. Typical Standards for Nonfat Dry Milk

	Regular				Instant[a]
	Spray-Dried		Roller-Dried		
Specification	Extra Grade (maximum values)	Standard Grade (maximum values)	Extra Grade (maximum values)	Standard Grade (maximum values)	Extra Grade (maximum values)
Milkfat, %	1.25	1.5	1.25	1.5	1.25
Moisture, %	4	5.0	4.0	5.0	4.5
Titratable acidity, %	0.15	0.17	0.15	0.17	0.15
Solubility index, ml	1.25[b]	2.0[b]	15.0	15.0	1.0
Standard plate count, CFU[c]/g	40,000	75,000	50,000	100,000	30,000
Scorched particles	Disc B (15.0 mg)	Disc C (22.5 mg)	Disc C (22.5 mg)	Disc D (32.5 mg)	Disc B (15.0 mg)

[a] Additional standards for instant include: coliform count, 10 CFU/g; % dispersibility, not less than 85 (modified Moats-Dabbah method). From (1); used with permission.

[b] Except products designated as "high-heat," which shall be not greater than 2.0 ml for Extra Grade and not greater than 2.5 ml for Standard Grade.

[c] Colony-forming units.

TABLE C-3. Specifications of Dry Whole Milk[a]

Standard	Extra Grade	Standard Grade
Milk fat, %	26–40	26–40
Moisture, % max.[b]	4.5	5.0
Solubility index		
Spray process, ml	1.0	1.5
Roller process, ml	15.0	15.0
Standard plate count, CFU[c]/g	≤50,000	≤100,000
Coliform count, CFU/g	≤10	≤10
Scorched particles		
Spray process	Disc B (15.0 mg)	Disc C (22.5 mg)
Roller process	Disc C (22.5 mg)	Disc D (32.5 mg)

[a] From (1); used with permission.

[b] As determined by weight of moisture on a milk solids-not-fat basis.

[c] Colony-forming units.

TABLE C-4. Composition and Standards for Caseins

Analysis	Acid Casein	Rennet Casein	Sodium Caseinate
Moisture, %	6.5–10.0	12.0	4.0–5.0
Protein, %	90.7	81.7	90–92
Fat, %	1.0–1.5	0.2–1.0	2.0
Lactose, %	0.2	0.1	0.2
Ash, %	1.6	7.5–8.0	3.0–4.0
pH	4.8	7.0	6.8
Standard plate count, CFU[a]/g	1,000–30,000	3,000–30,000	500–30,000
Coliforms	Negative	Negative	Negative
Listeria	Negative	Negative	Negative
Salmonella	Negative	Negative	Negative

[a] Colony-forming units.

TABLE C-5. Standards for Reduced-Lactose Whey, Reduced-Mineral Whey, and Lactose

Characteristic	Whey		Lactose	
	Reduced-Lactose	Reduced-Mineral	Edible Grade	Refined/USP Grade
Moisture, %	3–4	3–4	0.5 max.	0.5 max.
Total moisture, %	···	···	5.5 max.	5.0 max.
Lactose (hydrated), %	···	···	99.0 min.	99.8 min.
Lactose (anhydrous), %	···	···	94.0 min.	95.0 min.
Protein, %	18–24	11–15	0.1 max.	0.1 max
Fat, %	1–4	0.5–1.8	···	···
Lactose, %	52–58	70–80	···	···
pH	6.2–7.0	6.2–7.0	4.5–7.5	6.0–7.0
Ash, %	11–12	1–7	0.3 max.	0.3 max.
Density			0.90–0.95	0.90–0.95
Color	Cream to light yellow	Cream to light yellow	Light yellow to white	White
Flavor	Normal whey, salty flavor	Normal whey flavor	Clean, slightly sweet	Clean, slightly sweet
Scorched particle, mg	7.5–15.0	7.5–15.0	···	···
Standard plate count	<50,000 CFU[a]/g	<50,000 CFU/g	<10,000 CFU	<5,000 CFU/g
Coliform count	<10 CFU/g	<10 CFU/g	<10 CFU/g	<10 CFU/g
Yeast	···	···	<20 CFU/g	<10 CFU/g
Mold	···	···	<20 CFU/g	<10 CFU/g
Salmonella	Negative	Negative	···	···
Listeria	Negative	Negative	···	···
Coagalase-positive *Staphylococcus*	Negative	Negative	···	···

[a] Colony-forming units.

TABLE C-6. Typical Standards for Dry Cream

Standard	Allowed
Fat, %	72 ± 2
Protein, %	7.6
Carbohydrate, %	12 (by difference)
Minerals, %	2
Moisture, % max.	3.0
Standard plate count, CFU[a]/g	<10,000
Coliform count, CFU/g	<10
Salmonella, per 25 g	Negative
Yeast and mold, CFU/g	<10
Coagulase + *Staphylococci*, per 25 g	Negative

[a] Colony-forming units.

TABLE C-7. Typical Standards for Butter and Spray-Dried Butter

Standard	Butter	Butter Powder
Fat, %	80.0, min.	72 ± 2
Moisture, % max.	17.3	4.0
Color	Light to yellow	Light yellow
Sieve analysis	...	100% through No.12 sieve
Flavor	Sweet cream	Sweet cream butter flavor
Salt, %	0–2%	...
Standard plate count, CFU[a]/g	<1,000	<30,000
Coliform count, CFU/g	<1 CFU/g	<10
E. coli count	Negative	Negative
Salmonella	Negative	Negative/25 g
Yeast and mold count, CFU/g	<10	<10
Ingredients	...	Butter, nonfat dry milk, tocopherol, vitamin C

[a] Colony-forming units.

TABLE C-8. Specifications for Spray-Dried Buttermilk[a]

	Spray-Dry Process		Roller Process	
	Extra Grade	Standard Grade	Extra Grade	Standard Grade
Moisture, % max.	4.0	5.0	4.0	5.0
Milkfat, % min.	4.5	4.5	4.5	4.5
Protein, % min.	30.0	30.0	30.0	30.0
Titratable acidity, %	0.10–0.18	0.10–0.20	0.10–0.18	0.10–0.20
Standard plate count, CFU[b]/g	≤50,000	≤200,000	≤50,000	≤200,000
Scorched particles	Disc B (15.0 mg)	Disc C (22.5 mg)	Disc C (22.5 mg)	Disc D (32.5 mg)
Solubility index, ml	1.25	2.0	15.0	15.0

[a] From (1); used with permission.
[b] Colony-forming units.

Reference

1. American Dairy Products Institute. 1990. Standards for Grades of Dry Milks Including Methods of Analysis. Bulletin 916, revised. The Institute, Chicago, IL.

APPENDIX D.
Processing Guidelines

Chocolate Milk

Composition

Component	Chocolate Milk	Low-Fat Chocolate Milk
Milkfat, %	3.4	1.0
Milk solids-not-fat (MSNF), %	8.3	8.7
Sucrose, %	5.5	5.5
Chocolate powder, %	1.3	1.3
Total solids, %	18.5	16.3

Formulation for 100-Gal Batch

Ingredients	Chocolate Milk	Low-Fat Chocolate Milk
Skim milk, lb	...	777
3.4% Milk, lb	796	...
Nonfat dry milk (NFDM), lb	7	10
36.5% Cream, lb	10	27
Chocolate-flavored cocoa powder, lb	12	12
Liquid sugar, lb	75	74
Total, lb	900	900

Manufacturing Procedure

Blend all liquid ingredients. Sift chocolate powder slowly into liquid blend and agitate at high speed for 15 min. Pasteurize at 82.2°C (180°F) for 16 sec. Homogenize at 1,000 psi, single stage. Cool to 1.6–3.3°C (35–38°F). Package and store at 1.6–4.4°C (35–40°F).

Microbiological and other standards

Standard plate count, <10,000 colony-forming units (CFU)/g
Coliform, <1 CFU/3 ml
Psychrotrophic, <1 CFU/10 ml
Shelf life, 14 days at 7.2°C (45°F)
Weight/gal, 8.95 lb

Cultured Skim Milk/Nonfat Buttermilk (From 1,2)

A. Bulk Starter Preparation

Composition: 0.1% milkfat, 10.3% MSNF, 10.4% total solids

Formulation

Ingredient	Pounds for 100-Gal Batch
NFDM	93.5
Water	778.5
Total	872

Manufacturing Procedure

Add NFDM to water through a powder funnel. Pasteurize at 83.8°C (185°F) for 30 min. Cool to 22.2°C (72°F) and add one can of frozen culture per 100–300 gal of bulk starter. Mix well and hold at room temperature (22.2°C) until pH is 4.5 or titratable acidity is 0.83–0.87%. Incubation time is of the order of 16–18 hr. Cool to 7.2°C (45°F) and use within four days.

B. Nonfat Buttermilk Formulation and Manufacture

Composition: 0.1% milkfat, 10.0% MSNF, 0.18% salt, 10.28% total solids

Formulation

Ingredient	Pounds for 100-Gal Batch
Skim milk	849.3
NFDM, low heat	12.9
Dairy salt	1.6
Total	864

Manufacturing Procedure

Blend NFDM and salt into skim milk using a powder funnel and mix thoroughly. Heat treat at 90.5°C (195°F) for 5 min. Cool to 22.2°C (72°F). Inoculate with 0.75% bulk starter manufactured in section A. Ripen to 0.8% titratable acidity or pH 4.5. Usually this takes 16–18 hr. Break the coagulum and cool to 1.6–4.4°C (35–40°F). Package and store at 1.6–4.4°C (35–40°F).

Microbiological and other standards
  Contaminants, <1 CFU/ml
  Coliform, <1 CFU/3 ml
  Shelf life, 21–30 days at 7.2°C (45°F)
  Weight/gal, 8.65 lb

Yogurt (1,2)

A. Bulk Starter Manufacture

Composition: 0.10% milkfat, 10.30% MSNF, 10.40% total solids

Formulation

Ingredients	Pounds for 100-Gal Batch
NFDM	93.5
Water	778.5
Frozen culture (yogurt)[a]	1 can (70 ml)
Total	872

[a] (To be added later)

Manufacturing procedure

Add dry milk to water using a powder funnel while agitating the blend. Pasteurize at 87.8°C (190°F) for 30 min. Cool to 44.4°C (112°F). Add 1 can of frozen yogurt culture. Mix well and hold at 43.3°C (110°F) until pH of 4.7 or 0.8% titratable acidity is reached. Development time: 3–4 hr. Cool to 4.4°C (40°F) and store no more than five days. Weight/gal = 8.72 lb.

B. Natural/Plain Lowfat Yogurt

Composition: 1.00% milkfat, 14.23% MSNF, 0.75% stabilizer, 15.98% total solids

Formulation

Ingredients	Pounds for 100-Gal Batch
Skim milk	795
40% cream	22
Nonfat dry milk	57
Stabilizer	7
Total	881

Manufacturing Procedure

Blend cream and skim milk. Blend the dry ingredients and add to the liquid through a powder funnel. Heat treat at 90.5°C (195°F) for 10 min. Homogenize at 57.2–62.7°C (135–145°F) and 1,500 psi. Cool to 44.4–45.6°C (112–114°F) and transfer to a cone vat. Add yogurt bulk starter at the rate of 5%. Mix thoroughly. Hold at 43.8°C (111°F) until pH 4.6 is reached. Development time is 2.5–3 hr. Cool to 22.2–23.8°C (72–75°F). Set Waukesha pump line valve at 20 psi. Package and hold in the cooler for 48 hr.

Specifications

pH 4.2–4.3; no wheying off; smooth body and clean flavor

Microbiological and other standards

Contaminants, <10 CFU/g
Coliforms, <1 CFU/g
Yeast and mold, <1 CFU/g
Shelf life, 30 days at 7.2°C (45°F)
Weight/gal, 8.81 lb

C. Stirred-Style Low-Fat Yogurt

Composition: 1.00% milkfat, 12.00% MSNF, 9.00% sugar, 0.70% stabilizer, 22.7% total solids

Formulation

Ingredients	Pounds for 100-Gal Batch
Skim milk	750
40% cream	25
NFDM, low heat	44
Sucrose	77
Stabilizer	7
Total	903

Manufacturing Procedure

Blend cream and skim milk in a mix tank. Blend dry ingredients and add to the liquid through a powder funnel and agitate continuously. Pasteurize at 90.5°C (195°F) for 20 min. Homogenize at 57.2–62.7°C (135–145°F) and 1,500 psi. Cool to 44.4–45.6°C (112–114°F) and transfer to a cone vat. Add yogurt bulk starter at a 5% level. Hold at 43.3–43.8°C (110–111°F) until pH is 4.6 or titratable acidity is 0.85%. Development time is 2.5–3.5 hr. Cool to 22.2°C (72°F). Set Waukesha pump line valve for 20 psi to pump to fruit feeder. Incorporate yogurt fruit through the fruit feeder at a 10–15% level or as recommended by yogurt fruit supplier. Store at 1.6–4.4°C (35–40°F) for 24–48 hr to achieve viscosity target.

Microbiological standards

Contaminants, <10 CFU/g
Coliforms, <1 CFU/g
Yeast and mold, <1 CFU/g
Shelf life, 30 days at 7.2°C (45°F)
Weight/gal, 9.03 lb

References

1. Chandan, R. C., and Shahani, K. M. 1993. Chapter 1 in: *Yogurt in Dairy Science and Technology Handbook*, Vol. 2, Product Manufacturing. Y. H. Hui, Ed. VCH Publishers, Inc. New York.
2. Chandan, R. C., and Shahani, K. M. 1995. Chapter 10 in: *Biotechnology*, Vol. 9, 2nd ed. G. Reed and T. W. Nagodawithana, Eds. VCH Publishers, Weinheim, Germany.

Glossary

Acid whey—Water and milk solids left after removal of curd in the manufacture of cottage and ricotta cheeses. Its pH is about 4.4–4.6.

Aflatoxin—A highly poisonous and carcinogenic substance produced by the mold *Aspergillus flavus*. If contaminated peanuts, corn, or cottonseed are fed to cows, the milk may contain aflatoxin.

Alkalinity of ash test—Test to determine whether neutralizers (Na_2CO_3, $NaHCO_3$, $NaOH$, etc.) were used to neutralize the cream used for manufacturing butter, buttermilk, and dry milk products. It involves titration of ash with standard acid.

Amino acid—An organic acid containing both an amino (NH_2) and an acidic (-COOH) group; the building blocks of proteins.

Amphoteric—Describing a compound such as a protein that has both positive and negative charges.

Antioxidants—Compounds that inhibit lipid oxidation.

Ash—The residue left when a substance is incinerated at a very high temperature.

Biological value—The proportion of absorbed protein that is retained in the body for maintenance and/or growth. It is expressed as a percentage.

Biologics—Biological products used in medicine.

Bloom—A dusty white appearance on the surface of chocolate caused by the formation of certain types of fat crystals.

Casein micelles—Large colloidal particles that are complexes of protein and salt ions, principally calcium and phosphorus.

Caseinates—Compounds derived by the interaction of alkali with **casein**, the major milk protein.

Cation—A positively charged ion.

Chemical score—The content of the most limiting amino acid, expressed as a percentage of the content of the same amino acid in egg protein.

Chocolate liquor—The solid mass obtained when ground, liquefied cacao bean kernels are cooled.

Chymosin—A milk-clotting enzyme found in rennet and now derived by biotechnology.

Coliform count—A group of intestinal tract microorganisms that, if present in food or water, usually indicates the contamination of that food or water with fecal matter.

Collets—Puffed second-generation snacks obtained by extruding corn meal.

Colloidal phase—The portion of milk containing dispersed particles ranging in diameter from 10^{-5} to 10^{-7} cm.

Colostrum—The first milk secreted by an animal just before and after the birth of its young.

Comminution—Reduction to minute particles.

Compound coatings—Coatings containing fats other than cocoa butter but similar to regular chocolate in melting properties.

Continuous phase—A liquid phase of an emulsion in which another immiscible liquid is dispersed (the dispersed phase).

Creaming—1) High-speed mixing of a plastic shortening containing sugar in order to incorporate air. 2) In an emulsion, the collection of the lighter phase in the upper part of the mixture (e.g., oil droplets on top of water).

Cyclone—A centrifugal device for separating materials (e.g., particles from air).

DATEM—Diacetyl tartaric acid esters of monoglycerides, an emulsifier.

Denaturation—The process that proteins undergo when subjected to certain physical or chemical treatments (e.g., heating) that cause disruption of the non-covalent bonds that maintain their secondary and tertiary structure. Denaturation causes profound changes in functional properties.

Density—Mass per unit volume.

Diacetyl—A chemical compound characterizing the flavor of butter, milkfat, and certain fermented dairy products.

Draw temperature—The temperature at which frozen dairy products are removed from the freezer for optimum body and texture and to facilitate packaging.

Electrical conductivity—The reciprocal of electrical resistance exhibited by a 1-cm cube of conductor (solution containing electrolytes).

Electrodialysis—A process that uses electric charge to separate substances in solution (in this case, removing minerals from whey or milk fractions).

Electrolytes—Positively (cation) and negatively (anion) charged ions.

Emulsifying salts—Salts involved in emulsification of fat and stabilization of the emulsion.

Emulsion—A homogeneous dispersion of two dissimilar immiscible liquid phases. If oil is dispersed in water, it is an oil-in-water (O/W) emulsion. If water is dispersed in oil, it is a water-in-oil (W/O) emulsion.

Enteropathogen—A disease-producing microorganism in the intestinal tract.

European cultured butter—Butter made from cream to which a lactic culture has been added. It contains no salt and has a characteristic diacetyl flavor.

Extended shelf life—Shelf life of 14–28 days at refrigerated temperature.

Fat mimetic—A fat-replacing ingredient based on protein, starch, other carbohydrates, or hydrocolloids that mimics the properties of fat.

Fatty acids—A group of chemical compounds characterized by a chain made up of carbon and hydrogen atoms and having a carboxylic acid (COOH) group at the end of the molecule. When they exist unattached to other compounds, they are called free fatty acids.

Feathering—The limited coagulation of milk proteins when cream is added to hot liquid (e.g., coffee), characterized by the appearance of small particles on the surface.

Frozen custard—An ice cream containing at least 1.4% egg yolk solids. Also known as French ice cream.

Functional foods—Foods that have been shown by clinical trials to promote health, prevent disease, or help in the treatment of disorders. Also known as *designer foods, pharmafoods, biofoods,* or *nutraceuticals.*

Galactose—A monosaccharide (simple sugar) liberated when lactose is hydrolyzed by the enzyme lactase.

Gelation—The process of gel formation, in which globular proteins act as gelling agents and provide desirable texture by holding a large quantity of water. Caused by heat denaturation or by pH changes, salt addition, or enzyme action.

Hazard analysis and critical control points (HACCP)—A system of steps established by the food industry to produce safe products through anticipation and prevention of problems.

High-ratio—Describing a cake that contains more sugar than flour.

Homogenization—A process for reducing the size of milkfat globules in milk. Upon undisturbed storage at 7°C, homogenized milk shows no visible cream separation.

Hydrocolloids—Gums; water-soluble or modified polysaccharides used for thickening and water binding.

Hydrolysis—Enzymatic cleavage of glyceride molecules or proteins to liberate their constituent fatty acids or amino acids, respectively.

Hydrolytic rancidity—A flavor defect associated with the activity of the enzyme lipase on milkfat, which liberates short-chain fatty acids with objectionable odors.

Hydrophilic—Attracted to water (polar).

Hygroscopic—Readily taking up and retaining moisture.

Intestinal mucosa—The mucous membrane lining the intestine.

Ion—An atom or group of atoms that carries a positive or negative electric charge.

Lactase—The enzyme that splits lactose (milk sugar) into glucose and galactose.

Lactones—Chemical compounds derived from the hydrolysis of hydroxy fatty acids. Constituents of the overall flavor of dairy products.

Lactose—Milk sugar, composed of glucose and galactose.

Lactose intolerance—Maldigestion of milk sugar (lactose) by certain individuals, who experience abdominal pain, bloating, and diarrhea following consumption of milk and dairy products containing lactose.

Lecithin—A phospholipid found in egg yolk and soybeans and also used as a food ingredient. It is a surfactant that can stabilize emulsions.

Lipase—An enzyme that hydrolyzes acyglycerols/glycerides.

Lipids—A class of compounds found in nature that are soluble in organic solvents such as ether or hexane. Examples are triacylglycerols, cholesterol, and vitamin A.

Lipophilic—Attracted to fat (nonpolar).

Low-acid thermal processing—Processing in which foods, other than alcoholic beverages, are made to have a finished equilibrium pH higher than 4.6 and water activity more than 0.85. Used with aseptic packaging.

Low-fat milk—Milk containing at least 8.25% solids-not-fat and with fat reduced to deliver not more than 3 g of milkfat per serving of 8 fl oz. Also termed *light milk*.

Lubricity—A desirable slippery sensation in the mouth imparted by fats.

Maillard browning—Nonenzymatic browning caused by the heat-induced reaction of the ε-group of lysine with a reducing sugar. The bioavailability of lysine is lost, and characteristic flavor changes occur.

Masa—Steeped and cooked corn dough (or flour) used in the manufacture of corn snacks.

Methyl ketones—Constituents of the flavor profile of dairy products, similar to lactones. Derived from β-keto acids.

Microbiota—The microscopic flora and fauna of a region.

Milk crumb—An intermediate ingredient of milk chocolate, consisting of a mixture of milk solids, sugar, and chocolate liquor.

Net protein utilization—Proportion of protein intake that is retained (calculated as biological value times digestibility).

Nonfat milk— Milk containing at least 8.25% solids-not-fat and with fat reduced to deliver not more than 0.5 g of milkfat per serving of 8 fl oz. Also termed *fat-free* or *skim milk*.

Oestrum/Estrus—Period of sexual receptivity (heat) in female mammals.

Oiling-off—Liberation of free fat from cheese (e.g., pizza cheese) as a result of exposure to high temperature.

Oxidation—A chemical reaction in which the double bond on a lipid molecule reacts with oxygen to produce a variety of chemical products. The consequences of this reaction are loss of nutritional value and formation of the off-flavors associated with rancidity.

Parturition—The act or process of giving birth.

Pasteurization—The process of heating milk and fluid dairy products to render them safe for human consumption by destroying disease-producing microorganisms. The process inactivates about 95% of the contaminating microorganisms.

Peroxide value—A number that indicates the level of peroxides in a fat or oil; a measure of the amount of oxidation of the fat or oil.

Phase inversion—The reversal of emulsion phases, changing the emulsion type, e.g., from the oil-in-water type to the water-in-oil type.

Phosphatase test—Commonly used test for confirming properly pasteurized milk and milk products. It measures residual phosphatase, which would have been inactivated by proper heat treatment.

Plastic cream—An oil-in-water emulsion (cream) containing at least 80% butterfat.

Polybrominated biphenyls (PBBs)—A class of environmental contaminants that may be found in dairy products.

Polychlorinated biphenyls (PCBs)—Environmental contaminants that can accumulate in the food chain and gain entry into milk via animal feed. May cause a variety of harmful effects, including damage to the reproductive cycle of animals and plants.

Prebiotics—Nondigestible food ingredients that improve the host's health by selectively stimulating the growth and/or activity of one or more bacteria in the colon.

Probiotics—Live microorganisms introduced into the gastrointestinal tract of humans or animals to improve the microbial balance or microbial metabolic activity.

Process cheese—A cheese product made by blending various natural cheeses in the presence of heat and emulsifying salts.

Prooxidants—Certain metal ions in dairy products that accelerate oxidative deterioration and the development of off-flavors.

Protease—An enzyme that attacks and hydrolyzes proteins.

Protein efficiency ratio—Gain in body weight divided by weight of protein consumed.

Proteolytic activity (proteolysis)—The enzymatic breakdown of the protein molecule to yield peptides (smaller fragments).

Psychrotrophic—Describing cold-tolerant microorganisms capable of growing at 4–15°C (39–59°F).

Pumped hams—Hams with added water containing curing salts.

Rancidity—An off-flavor caused by oxidation of fat or by the release of flavorful fatty acids from a triacylglycerol/triglyceride.

Reducing sugar—A sugar molecule containing free aldehyde groups that are available to react with a free amino group of protein, peptide, or amino acid.

Reformed hams—Hams processed by cooking pieces together to form a single piece.

Refractive index—A physical property of a substance that relates to how light is refracted from the material. Usually used to indirectly measure some other property such as concentration.

Rennet—An extract of the fourth stomach of a calf, kid, or lamb, used in the coagulation of milk during cheesemaking.

Rennin—A milk coagulatory enzyme found in the gastric juice of the fourth stomach of calves.

Serum phase—The watery portion of a fluid; the portion of milk without fat globules and casein micelles.

Somatic cell count—Count of the mixture of dead epithelial cells and leukocytes that migrate into milk from the udder.

Specific gravity—Ratio of the density of a product and the density of water at the same temperature.

Specific heat—Number of calories required to raise the temperature of 1 g of a substance by 1 degree C.

Specifications—A set of chemical or physical quality requirements that a product must meet before it is acceptable.

Standard of identity—A legal standard, maintained by the FDA, that defines a food's minimum quality, required and permitted ingredients, and processing requirements, if any. Applies to a limited number of staple foods.

Standardization—A step in dairy processing in which milkfat and/or solids-not-fat are made to conform to certain specifications by removal, addition, or concentration of milkfat.

Surface tension—Forces causing a reduction in surface area, which is a characteristic property of a liquid.

Surimi—A processed fish product made from mild-flavored white fish that is given the shape, texture, and flavor of lobster, crab, scallops, or shrimp.

Sweet whey—Water and milk solids left after removal of curd in the manufacture of Cheddar, Swiss, and mozzarella cheeses. Its pH is about 5.5–6.0.

Syneresis—The separation of liquid from a gel; weeping.

Thermoduric bacteria—Bacteria that survive pasteurization.

Titratable acidity—Test used for determining milk quality and for monitoring the progress of fermentation in cheese and fermented milks. It measures the amount of alkali required to neutralize the components of a given quantity of milk and milk products and is expressed as percent lactic acid.

Triacylglycerols—Compounds of glycerol with three fatty acids attached. Neutral fats are composed of mono-, di-, and triacylglycerols (triglycerides).

Trimethylamines—Compounds responsible for a fishy odor in dairy products.

Ultrafiltration—A process that uses a semipermeable membrane to separate fractions based on molecular size.

Ultrahigh temperature—A temperature of 135–150°C (275–302°F), used with a holding period of 4–15 sec. This sterilizes milk to permit storage at ambient temperature.

Ultrapasteurization—The process of pasteurizing fluid dairy products by heating to 125–137.8°C (257–280°F) with a holding time of 2–5 sec to kill all pathogenic organisms, permitting storage at refrigerated temperature for an extended period.

Viscosity—Resistance to flow. A measure of the friction between molecules as they slide past one another.

Whey—The watery liquid remaining after the curd is formed in the manufacture of fermented or acidified dairy products.

Whey protein nitrogen test—Test used as a measure of the degree of heat received during processing of dry milk and whey protein concentrates. It is correlated with protein denaturation and with certain functional characteristics of dairy concentrates.

Index